METHUSELAH'S ZOO

METHUSELAH'S ZOO

**What Nature Can Teach Us
about Living Longer, Healthier Lives**

STEVEN N. AUSTAD

**The MIT Press
Cambridge, Massachusetts
London, England**

The MIT Press would like to thank the anonymous peer reviewers who provided comments on drafts of this book. The generous work of academic experts is essential for establishing the authority and quality of our publications. We acknowledge with gratitude the contributions of these otherwise uncredited readers.

This book was set in Adobe Garamond Pro and Berthold Akzidenz Grotesk by Westchester Publishing Services. Printed and bound in the United States of America.

Library of Congress Cataloging-in-Publication Data

Names: Austad, Steven N., 1946- author.
Title: Methuselah's zoo : what nature can teach us about living longer,
    healthier lives / Steven N. Austad.
Description: Cambridge, Massachusetts : The MIT Press, [2022] | Includes
    bibliographical references and index. | Summary: "A natural history of
    longevity in a wide variety of species along with an exploration of what
    we can learn from other species to preserve and extend human health"—
    Provided by publisher.
Identifiers: LCCN 2021049887 | ISBN 9780262047098 (hardcover)
Subjects: LCSH: Animals—Longevity. | Longevity. | Aging—Prevention.
Classification: LCC QP85 .A97 2022 | DDC 591.4—dc23/eng/20211221
LC record available at https://lccn.loc.gov/2021049887

10 9 8 7 6 5 4 3 2 1

To Veronika, who understands

Contents

Preface

It was opossum number 9 that hooked me. Working from a biological research station on the central Venezuelan savanna, my friend and colleague Mel Sunquist and I had begun a project to investigate how nutrition might affect whether opossums bore mostly male versus mostly female pups. I had first tagged number 9 when she was no bigger than a honeybee, hairless and eyes still shut, suckling in her mother's pouch. More than a year later, I caught her in a cage trap. Now full adult size, I fitted her with a radio collar and then recaptured her every month for the rest of her life. At fifteen months old, she looked in perfect health, a vigorous opossum mother with eight healthy pouch young. But three months later, I was shocked to find that she had developed cataracts in both eyes. She had lost weight. The muscle along her flank had noticeably shrunk. When I released her, she wobbled along even more sluggishly than most opossums. Less than a month later, she was dead. How could she have aged so dramatically in three short months?

Up to then, I hadn't thought much about it, but I assumed that opossums, which are about the size of a house cat, would age pretty much like a house cat. House cats—and I had had plenty of them in my pet-filled life—remain vigorous and healthy through at least their first decade or decade and a half of life. Yet here was a year-and-a-half-old opossum who looked and acted ancient. Years later, by which time I had tracked more than a hundred wild opossums from birth to death—few of them surviving to age two and none surviving even as long as age three—I dug into the sparse scientific literature to see what it said about their longevity. Wild opossums live at least seven years, that literature told me. How could what I saw—when I had followed so many individuals almost from the day they were born to the die they died—be so different from what the scientific literature claimed? Eventually, I tracked down the reason. Someone, years previously, had measured a number of opossum skulls in the Smithsonian Natural History Museum's collection. Making assumptions about opossum's reputed continued growth throughout life, they had calculated that the biggest skull must represent an opossum at least seven years old.

By now though, I knew wild opossums well, as well as anyone in the world. Having moved from Venezuela back to the States, I had taken a position as an assistant professor at Harvard University, where I had a large opossum longevity study underway. North American opossums also seemed elderly by no more than two years of age. But one of my radio-collared animals had grown up in an area near an open dumpster into which a local restaurant disposed of its waste food. Scavenging that food night after night, she was nearly twice the size of any other of my study opossums the same age. Aha! Now I understood how that unusually large skull in the Smithsonian collection could have misled someone.

Not long after my encounter with opossum number 9 in the wilds of Venezuela, I lost interest in whether nutrition affected whether animals had mostly male or mostly female offspring. What was going on with aging? Why did some animal species age fast and die young and others, outwardly very similar, age slowly and die old? Why couldn't nature—which routinely performed the almost miraculous transformation of a fertilized egg into a healthy adult frog or fish or ferret—do the seemingly much easier task of maintaining that adult's health? And could we understand more about this mysterious process of aging if we knew more about the real lives of animals in their natural habitat? The misinformation I had uncovered about opossum longevity made me curious about misinformation about other species, including our own.

That was almost forty years ago. Over the intervening decades, I have been steeped in research on why and how animals age. Although I kept my interest in how long animals could live in nature, most of my research and that of my colleagues trying to understand the biology of aging was carried out with laboratory species that were demonstrably unsuccessful at coping the aging process. They all lived fast and died young. With all the wonders of twenty-first-century cell and molecular biology, we were making spectacular progress at understanding the way these laboratory species aged. We were discovering many ways to slow the aging process in these species. But humans already age much more slowly than any of our laboratory animals. Were we learning anything relevant to keeping people healthy longer from the study of animals that lived a few weeks or a few months or a few years? Might the laboratory of nature, in which some species had evolved to be much more successful than humans at staving off the depredations of aging, have something to teach us that we would never learn from our laboratory bestiary of tiny worms, fruit flies, and domesticated mice?

That question is the genesis of this book. Combining my own passion for natural history with my professional interest in finding new ways to extend human health, I wanted to explore—in more depth and with more attention to separating fact from speculation from wishful thinking than anyone had ever done before—the details of exceptional longevity in the wild. It is only by having the actual facts that nature might lead us forward.

Without the generous help of many people, this book would not have been possible. I received no end of expert advice on species of their special expertise from biologists I have met and whose research I have admired over the years. These people include Carol Boggs, Bert Hölldobler, Laurent Keller, and Barbara Thorne for ants, termites, and other insects; Emma Teeling and Jerry Wilkinson for bats; Thane Wibbels for turtles and tortoises; and Lindsey Hazley for tuataras. Howard Snell gave me insight on all Galápagos species. Ken Dial, Geoff Hill, and Bob Ricklefs offered me their expertise on birds over the years. Stan Braude and Shelly Buffenstein provided information on naked mole-rats. For the latest information on elephants, I am grateful to Daniella Chusyd, Mirkka Lahdenperä, and Phyllis Lee. For chimpanzees, Steve Ross and Melissa Emery Thompson provided invaluable input. For information on aging fishes and sharks, I'm grateful to Allen Hia Andrews, Greg Cailliet, and Steve Campana. For dolphins and whales, information and opinions provided by Aleta Hohn,

Janet Mann, Todd Robeck, Peter Tyack, and Randy Wells were particularly helpful. I am grateful to Chris Richardson and Iain Ridgway for introducing me to bivalve mollusks, the longest-lived group of animals. The people keeping records on thousands of captive animals deserve special credit as well. I am particularly indebted to Beth Autin and Melody Brooks of the San Diego Zoo, Lindsay Hazley of New Zealand's Southland Museum and Art Gallery, Debbie Johnson of the Brookfield Zoo, Steve Ross of the Lincoln Park Zoo, and Joann Watson of the Houston Zoo. Any errors, of course, are entirely my own.

In the aging field, per se, I thank João Pedro de Magalhães, who took several decades of my own record keeping, added it to his own, systematized it all, keeps it up to date, and has made it public and searchable via his outstanding website on animal longevity, AnAge (https://genomics.senes cence.info/species). For their generosity with ideas, insights, and friendship over the years, I'm particularly thankful to Nir Barzilai, Tuck Finch, Keyt Fischer, Jim Kirkland, George Martin, Richard Miller, Jay Olshansky, Arlan Richardson, Felipe Sierra, Dick Sprott, and Gary Ruvkun, my partner in crime at the Woods Hole summer course on the molecular biology of aging. I also thank many of the people above for reading large parts of this book as it was coming together, in particular Gary Dodson, Jessica Hoffman, and Veronika Kiklevich. Rick Balkin made valuable comments on the whole thing. I would also like to thank my agent, Anthony Arnove, and Bob Prior, my editor at the MIT Press, without whom this book would have been stillborn. Finally, for putting up with my disappearance for long hours while I was writing or away in the field, I thank my wife, Veronika Kiklevich, and daughters, Marika and Molly.

1 DOCTOR DUNNET'S FULMAR

I am looking at two photographs of the Scottish ornithologist George Dunnet. In the first, he is a slender, bright-eyed, twenty-three-year-old man with black, curly hair. Cupped between his hands is a bird. To the uninitiated, the bird might be indistinguishable from a seagull. Aficionados though will recognize it as a northern fulmar (*Fulmarus glacialis*), a relative of the albatross that can be found during its breeding season nesting along the coastal cliffs and islands of the North Atlantic Ocean. When not breeding, it spends its time far from land, soaring over the open sea. The year is 1951, and Dunnet has just begun a study of an island colony of northern fulmars that he will continue for the rest of his life.

The second picture was taken thirty-five years later (figure 1.1). The man, now age fifty-eight, has changed fairly dramatically over that time, as we all do. He is stouter, grayer, a bit more weather-beaten. He certainly no longer looks like he could carelessly bound among the cliffs of Eynhallow Island in the Orkneys, his study site. He is holding a bird in this second photo as well. Yes, it is the same bird, which doesn't appear to have changed at all. Not only is the bird still young looking (to a human eye), but Dunnet reported that his bird was still as reproductively active as ever—something I suspect you couldn't say of Dunnet himself. The bird was still knocking out one chick per year, as it had been for decades and as it would continue to do even after Dunnet's death some nine years later. The bird was also still capable of working its tail feathers off. In order to continue being reproductively successful, a northern fulmar must make repeated foraging trips, some covering as much as four thousand miles, over the ocean before

Figure 1.1
Ornithologist George Dunnet in 1951 at age twenty-three and 1986 at age fifty-eight. The same bird is shown in both photos. Dunnet died in 1995. The bird was last seen the year after his death.
Source: Photo courtesy of the Outer Hebrides Natural History Society.

returning with a stomach laden with fish, squid, and shrimp to nourish its growing chick.

A number of bird species live a long time, some even longer than this fulmar, as we shall see. Perhaps even more astonishing than their long lives, though, is that they continue at an advanced age to meet the enormous energy demands that their lives require, such as those long overseas flights. Birds in nature somehow seem to remain physically fit to the very end of their lives. Wouldn't it be nice if people could do something similar?

NATURAL LIFE SPANS

The subject of this book is long life among animals in the natural world. Long life in nature is a rare trait but is broadly distributed across the animal kingdom. The book is also about how and where and by whom long life is achieved and what we might learn about keeping ourselves healthy longer from understanding how their biology allows these species such long life.

Nature imposes two general impediments to long life—impediments that most species are unable to overcome. One of these you might call *environmental hazards*—external threats to survival such as predators, famine, storms, drought, poisons, pollutants, accidents, or infectious diseases. We can estimate the impact of such hazards on longevity by comparing animals' longevity in the wild with their longevity in zoos, households, or laboratories, where we pamper and protect them from these hazards.

Consider the humble house mouse (*Mus musculus*) in this regard. In nature, its life expectancy is three to four months. The domesticated form of the house mouse is the laboratory mouse, a mainstay of medical research. The lab mouse is the poodle to the wild house mouse's wolf. In a well-managed laboratory colony, mice live two to three years—some eight to twelve times longer than in the wild. If you are a small, nearly defenseless mouse, nature fairly bristles with danger. Nature, however, provides serious challenges even for animals much better able to avoid them than a mouse. The raven, for instance, could seemingly fly away from many types of danger. Yet even ravens live about three times longer when protected in captivity than they do in the wild. Animals that achieve long life in the wild must have an exceptional ability to avoid or overcome environmental hazards.

The other impediment to long life is internal. We call this hazard *aging*. Aging—which in this book means not the simple passage of time but the progressive deterioration over time of bodily functions and defenses along with increasing susceptibility to diseases that bedevils us all—in this sense is nearly universal.

Aging occurs at vastly different rates in different animal species, though, as we can tell by comparing our own aging to that of our pets. Relative to dogs or cats, we humans age—that is, deteriorate—slowly. Yet their aging resembles ours in many ways. With time, they lose strength and endurance. Their fur grays. They get cataracts and arthritis. Their hearing fades. They more rapidly begin to suffer from internal errors that produce organ failure. I watched this happen to my first pet dog, Spot. My family got Spot when he and I were both young.

I had just started school. By the time I began high school, Spot was beginning to slow down. By the time I went away to college, Spot had passed on. Similar changes happen to us, of course, but they happen many times slower. Folk wisdom tells us that one human year is equal to seven dog years.

However, we ourselves age quickly compared to some other species. Aging slowly, in addition to being able to avoid environmental dangers, is an inescapable necessity for long life in nature. Better yet, of course, would be not aging at all.

A book appeared in 1992 entitled *Sharks Don't Get Cancer*.[1] The title makes quite a claim—a bogus claim, to be sure, but an eye-catching one. Sharks do get cancer. Mice get cancer, and dogs and cats and elephants get cancer. Long-lived parrots and tortoises get cancer. Every nearly species of mammal, bird, reptile, and fish that has been looked at in any depth gets cancer. This is because cancer is caused primarily by aging and almost all species age if their environment doesn't kill them first.[2]

Cancer, as we all know, is cell division run amok. Most (not all though) tissues of our bodies need a continuous supply of new cells to replace worn, damaged, and discarded ones. We get those new cells from existing cells that grow and divide to produce them. The lining of your intestines, for instance, is completely replaced by new cells every two to four days. Your skin cells are replaced in a month, and your red blood cells in four months. In fact, to keep up with all this cell replacement your body, as you sit there reading this, is manufacturing about two thousand miles of new DNA per second.

Let me repeat that in case it slipped by you. Your body manufactures a rather miraculous two thousand miles of new DNA per second just to replace the DNA it lost in discarded cells! Every time a cell divides, however, and synthesizes all that new DNA for the new cell, it is susceptible to DNA copying errors called *mutations*. Cancer occurs when enough critical mutations accumulate in a cell's DNA to make it lose control of its tightly regulated replication schedule. Sometime the loss of control is minor and leads only to lumps and bumps where we previously had none. Sometimes it fatally progresses to complete and unlimited loss of control, where cells keep on dividing without limit, eventually invading surrounding tissue, and spreading through the bloodstream to other parts of the body.

Ironically, considering that I am writing about long life, even normal cells (if removed from your body and grown in a culture dish) will divide only a certain number of times before permanently stopping. Cancer cells, by contrast, are immortal. For instance, cells from the cervical cancer of a woman named Henrietta Lacks were first grown in a laboratory dish in 1951 and have continued dividing ever since. The total weight of cells grown from Lacks's original biopsy in medical laboratories is now said to have exceeded twenty tons. These HeLa cells, after the first letters of her two names, were used in the 1950s to grow polio virus in the laboratory, leading directly to the Salk polio vaccine in 1954. Still, the corrupt vitality of cancer cells is bad for a living body.

Cancer ultimately results from damage to a single, original, "renegade" cell, as researcher Robert Weinberg memorably phrased it, which through a series of mutations has become a raging replication machine. The body of an older animal will be composed of cells that are the products of many more divisions than the body of a younger animal. Thus, older animals' cells will have accumulated many more mutations because when cells divide, they are most vulnerable to those DNA mutational copying errors. The more divisions, the more mutations, eventually leading to the loss of replicative control.

My statement that cancer is caused primarily by aging may be surprising given how frequently you hear about childhood cancer. Childhood cancer, without question, is a terrible tragedy when it occurs, but it is rarer than you might guess from the attention it receives. Fewer than one in two hundred cancer deaths occur in people under twenty-five years old. After that, the death rate from cancer doubles about every eight years, so that by age eighty-five, you are more than three hundred times more likely to die from cancer than you were before age twenty-five.[3] Cancer incidence progressively increases with age in all species for which we have information.

Even though almost all species age and even though all species that rely on cell division to repair themselves get cancer, those facts together do not dictate that all species are similarly susceptible to cancer. We know of some species that are remarkably resistant to cancer, in fact. That sort of resistance, along with similar resistance to aging generally, turns out to be a large part of the story about why some species live for an exceptionally long time.

WHY DO WE AGE?

Why nearly every creature ages, as contrasted with maintaining youthful health forever, is one of the enduring puzzles of biology. Evolutionary biologist George Williams summed up this puzzle succinctly by noting that evolution is remarkably adept at creating from a single fertilized egg a healthy trillion-celled young adult dog or dove or dolphin. Yet it seems incapable of performing the seemingly much easier task of maintaining the health of those adults once they have been created.

Equally puzzling is why some species age rapidly, deteriorating over the course of days or weeks, whereas for other species, it takes years, decades, or in some cases even centuries. Notice that I said "nearly every" living creature undergoes aging. A few seemingly do not. Those species will be of special interest as we ponder whether there are keys to slowing human aging that nature in all its cleverness has provided.

Actually, we have figured out in general terms why nature seems incapable of stopping aging. We also understand in general terms why aging occurs rapidly in some species but slowly in others. Those explanations will emerge as we investigate throughout the rest of this book how exceptionally long life is distributed within the animal kingdom.

A long life in nature, we have established, requires overcoming both external and internal hazards. Failing to overcome one as compared with the other has dramatically different consequences, though. Failure to overcome environmental hazards may lead to a short life, but it will be a largely healthy life. Consider once again the humble house mouse. In the wild, it dies of cold, predation, wounds, stress, disease, exhaustion, and starvation long before aging begins to take a substantial toll on its body. To the day of its death, its muscles remain strong, its senses sharp, its mind clear. By the time a mouse dies a "natural" death in the laboratory where it has been protected from environmental hazards, it dies from internal failure. And that death looks very different. In the lab, aged mice are likely to be blind, deaf, weak, arthritic, paralytic, and tumor-ridden by the time they die.

Nature has supplied many species—although not the mouse—with a variety of clever stratagems for avoiding or defeating external threats. However, doing so without also dealing with the internal threat of aging may be a fool's success. Life may be long, but at least in its latter stages, it is likely to be miserable due to the ravages of aging. This is the situation that humans currently face.

During the twentieth century, life expectancy in the economically developed countries of the world increased by about thirty years.[4] We haven't changed the rate at which we biologically age, however. We have simply made our environment more hospitable with better and better public health practices and increasingly sophisticated medical care. Before 1900, we were more like mice in the wild, dying of accidents or infections for the most part before significant debilitation crept in. Today, we are more like mice in the laboratory. Deaths in early life are rare. In fact, in the United States only about one person in twenty now dies before age fifty. The majority of us die from diseases of aging like cancer, heart disease, Alzheimer's disease, stroke, and kidney or lung failure. Even if we avoid these fatal diseases, our latter years can be marked by chronic pain, vision and hearing loss, and physical frailty. Our human life span has increased faster than our health span. If that trend continues, a societal disaster awaits. Health-care systems may well collapse under the weight of the frail and ill elderly unless we can find a way to treat the aging process as we treat diseases. Some of the species discussed in this book, already more successful than we are at staving off aging, may be able to point us toward scientific approaches to do exactly that. Others may have less to teach us about human aging, but they are interesting in their own right. I personally find exceptionally long-lived animals intrinsically interesting in the same way I find birds with spectacular plumage or mammals capable of extraordinary athletic feats interesting.

As I say, some species *are* successful at overcoming both external and internal hazards. Consequently, they live exceptionally long as well as exceptionally healthy lives. These are the species of what I call Methuselah's Zoo. They are the species on which we will focus—the species from which we may learn. Methuselah, as you may recall, is the longest-lived person mentioned among all the "begats" of the biblical patriarchs in the book of Genesis. Methuselah, the Bible claims, lived 969 years. Maybe equally remarkable, he reputedly fathered his first child, a son, at the age of 187 years. Perhaps in those days, adolescence was even more awkwardly extended than it is today. We will dig into the details of human longevity eventually, after we explore the biology of other exceptionally long-lived species.

First, however, we need to define *long life* so that we know it when we see it. That isn't as straightforward as you might think.

WHAT IS A LONG LIFE?

Aristotle made the first attempt to establish patterns of longevity among different species some 2,500 years ago. From a modern perspective, it was a brilliant analysis except for one thing. Aristotle knew almost nothing about how long different animals actually lived and what he thought he knew was often wrong. For instance, he thought that mollusks—which include squid, snails, and clams—lived for only a year (he was off a few hundred-fold on that one) and that the longest-lived animals were those that had feet (humans and elephants, specifically). He was wrong about that, too. Remarkably enough—despite this primitive, often erroneous knowledge about species longevity—he noticed two major patterns that were very real.

First, he noticed that while many plants live only a year—we call these appropriately enough *annuals*—some plants live far longer than any animal. Specifically, Aristotle was thinking of trees. Of course, we now know, thanks to their habit of producing annual growth rings, that many tree species live for centuries, and some even for millennia. I remember vividly as a child visiting a redwood forest in northern California, and being enthralled by a display of a giant slab of wood from an ancient tree. Its growth rings were marked with various historical dates—the birth of Christ, the fall of the Roman empire, the signing of the Magna Carta, Columbus's first voyage to America, the beginning of the Civil War, and so on. The existence of annual tree growth rings, at least for those trees that grow in seasonal environments, has allowed us to learn a great deal about tree longevity.

However, I won't be discussing the long lives of trees or other vegetation in this book because the meaning of the longevity of trees is often difficult to grasp for the following reason. One of the oldest-known trees in the world is named Old Tjikko (figure 1.2) (exceptionally old trees like exceptionally old animals have individual names, it seems), a sixteen-foot-tall Norway

Figure 1.2
Old Tjikko, the world's oldest known tree in all its scraggly glory. The text explains why Old Tjikko, despite its age, is unlikely to teach us much about healthy aging.
Source: Photo courtesy of Petter Rybäck.

spruce (*Picea abies*) that apparently got lost and ended up on a mountain in Sweden. Old Tjikko was 9,558 years old at last count. Yet you could easily wrap your arms around it.

At this point, something should seem wrong to you. Almost ten thousand years old? Only sixteen feet tall, and you can wrap your arms around it? For comparison, the General Sherman tree, the oldest living giant sequoia tree in California, is 2,500 years old, a whippersnapper by comparison, but is 275 feet tall and thirty-six feet in diameter! It would take about sixteen professional basketball players holding hands to wrap their arms around it.

The difference between these trees is that the General Sherman is what we would think of as an individual tree. Old Tjikko is something else. The visible part of Old Tjikko—the trunk, branches, needles, cones, and all—is not that old at all, maybe only a few hundred years. Norway spruce are clonal, as are many other trees, meaning the old parts are underground—a system of roots that may spread over many acres. That root system sends up shoots, maybe dozens or even hundreds of them, which we see as individual trees, but in fact, they are genetically identical cloned stems sprouting from the same root system. Old Tjikko is that kind of tree. That "tree" may die, but the root system lives on, continually sending up other shoots—that is, other "trees." If you cut down Old Tjikko and counted the growth rings in its trunk, you'd think it was only a few hundred years old. And you would be right—for that particular stem but not for that particular plant.

Equally confusing, if you cut off one of Old Tjikko's branches and stick it in the ground elsewhere, it might sprout and start its own new root system. How old is that plant now? Is it the age of the root system from which it originally sprang or the age of the shoot from which it was cut? It is the existing root system of Old Tjikko, which has been carbon-14 dated to almost ten thousand years old. North America's quaking aspen is another "tree" of this type. A single genetic individual aspen may cover many acres, and its root system may be thousands of years old.

I am not saying that this isn't interesting from a scientific perspective, but it is a lot different than the longevity of individual animals such as a mouse, a man, or a monarch butterfly. It is of interest from the standpoint of how vegetative reproduction and the longevity of individual plant parts relate to one another. However, I'm most interested in how animals solve the problems of external and internal threats. I also feel more comfortable with animals that don't make me have to scratch my head to figure out whether they are really individuals and if those individuals are really old or not. So we will stick with animals in this book—animals that are clear individuals, like dogs and bees, squid, clams, and bats. We will avoid animals like those that build coral reefs. They are more like Old Tjikko in that defining their longevity will keep us tossing and turning at night.

The other pattern that Aristotle noticed was that large animals generally live longer than small ones. This turns out to be one of the most widespread and reliable patterns in nature. Among mammals this is pretty intuitive, something I think that we all expect from casual observations of animals in our daily lives. You wouldn't be surprised to learn that whales live longer than horses. Horses, in turn, live longer than dogs, which live longer than mice, and so on. Maybe not so expected, the same pattern holds among birds. A gull lives longer than a blackbird, which lives longer than a sparrow. Among reptiles, you find the same pattern, too, as you do among amphibians, even among clams. Note that these are general patterns—not rules. There are plenty of exceptions. Some species are exceptionally long-lived *for their size*. Humans are one of these. Others are exceptionally short-lived for their size. Mice are one of these. To jump ahead a bit, *Tyrannosaurus rex* is quite the exception to the general size-longevity rule as well. So are our primate relatives. Some other species that you might not expect are particularly dramatic exceptions.

The general size effect presents us with a problem for defining exactly now to recognize a long life. Do we focus on absolute longevity (the actual number of years of life) or relative longevity (the years of life compared with other animals of the same size)? Should a naked mole-rat, which is the size of a mouse but lives more than ten times as long, be considered a member of Methuselah's Zoo? It lives almost forty years, but that is less than a monkey, not to mention a human.

Actually, it does makes sense to consider size in defining Methuselah's Zoo. Here's why. Biological time flies by faster for small animals compared with large animals. Compared to a human or a horse, a mouse's engine is revved much higher. That is, each of a mouse's cells has a metabolic rate—that is, it burns energy—some fifteen times faster than the cells from a horse. For many years, metabolic rate has been thought to play a major role in aging.

Size isn't related only to metabolic rate though. Smaller animals live faster in many ways. They grow to adult size and begin reproducing faster. Their hearts beat faster. They breathe faster. Their muscles contract faster. Food churns through their guts faster; their blood and skin cells are replaced faster; their kidneys process waste faster. You get the idea. Physiological time runs faster for the smaller animals.

Small size also imposes more risks from environmental hazards. Smaller animals have more predator species capable of dining on them. Because of their relatively high energy demand compared to larger species, they must eat and drink more often and are therefore more vulnerable to starvation and dehydration than larger animals. Also, for the same reason that a small ice cube will freeze or melt more quickly than a large block of ice, small animals heat up and cool off quickly, so they are more vulnerable to temperature extremes.

Species that overcome all the limitations imposed by small size are worthy of our interest because we may be able to learn something valuable about ourselves by understanding how they overcome their challenges. But we need a measure—a way to compare different species that accounts for their different sizes.

In 1991, my former graduate student Keyt Fischer and I came up with a quick and easy, albeit crude, measure to do so.[5] We called it the *longevity quotient*, or LQ for short. It works like this. If you assemble the world record longevities of hundreds of mammal species ranging in size from shrews to elephants and fiddle with some simple mathematics, you can calculate how long an average mammal of any given size is expected to live, assuming it lives in a zoo or your house, where it is protected from environmental hazards. This is an important point. A species LQ is its longevity measured under protected conditions for the simple reason that those are the conditions for which longevity data are available for hundreds of species. Thank-you, zoos of the world.

Captive mammal longevity quotient will be our standard for all animals. By definition, the size-based longevity of an *average* mammal is an LQ of 1.0. Dogs, it turns out, have an LQ of just about exactly 1.0. From the standpoint of longevity, they are very much average mammals. If a species lives twice as long as an average mammal of the same size, it has an LQ of 2.0. If it lives half as long, it has an LQ of 0.5. A mouse has an LQ of about 0.7, meaning it lives 70 percent as long as an average mouse-size mammal under protected conditions. A wild mouse has an LQ of about 0.17, the difference due to living under hazardous natural conditions. People are long-lived mammals both in absolute terms and for our body size. Human LQ has various subtleties that we will consider later. For now, let's just say it is large, although not as large as some mammals. Once we stray outside the mammals, though, human longevity, either in absolute or relative terms, looks less and less impressive.

ACCURACY OF ANIMAL AGES

One final note before we take to the air, where we will begin our tour of longevity in the animal kingdom. I want to say a word about the accuracy of longevity records. It wasn't only ancients like Aristotle who made erroneous claims about how long animals live. Before widespread record keeping became common in the twentieth century, there was little reliable information on how long animals—or even people—lived. Documentation of birth dates for people much less for other animals is a pretty modern human phenomenon.

Without such documentation, a Thomas Parr becomes famous, for instance. Parr was a country bumpkin in sixteenth-century England who became a celebrity because he claimed without a shred of evidence to be 152 years old and people believed him in that less skeptical age. He died shortly after being dragged to London to meet King Charles I and was given a hero's burial in Westminster Abbey alongside England's most famous historical celebrities. You can find his grave in the cathedral floor even today. Exaggerating your age can make you famous.

Age exaggeration, for people or animals, has not gone away. I read bogus claims even in the respectable press almost weekly. I wrote about this extensively in my 1997 book, *Why We Age*. We will revisit some of these and more recent claims in the last chapter, which is about exceptional human longevity. For the present, suffice it to note that among professional demographers of aging, it is well known that the best way to live an exceptionally long life is to be born in a small village in a remote, preferably mountainous region, do hard physical labor your whole life, have a strong web of social support, and most important, live where illiteracy is common and reliable birth records are not.

Animal age exaggeration may possibly be even more rampant because birth dates of animals are seldom systematically recorded. Every time a giant tortoise dies in a zoo anywhere in the world, it seems that zoo personnel report that she was at least 175 years old and was donated to the zoo by Charles Darwin during his return voyage from the Galápagos Islands. This claim seemingly does not depend on whether the tortoise is actually a

Galápagos tortoise or whether the zoo is located somewhere that Darwin is known to have visited. For animals that live exceptionally long, we often have no better records than word of mouth passed down several generations or records scribbled on faded note cards buried in zoo files a century ago. That age exaggeration would be common isn't too surprising as there is often favorable publicity and financial gain to be had if you can brag about your particularly old animal in a zoo or as a pet. It becomes a curiosity and attraction rather than just another animal. It also becomes a quiet statement that the zoo or person in question takes particularly good care of its animals.

Sometimes the stories behind these bogus longevity records are as interesting as the actual records themselves. Stanley Flower, an early compiler of animal longevity records, noticed that a famous zoo elephant named Little Princess, who was reputed to have lived 157 years, managed to change from an African elephant to an Asian elephant sometime during her life, a feat considerably more impressive than living 157 years.

One of my personal favorites among stories about long-living animals is a female blue and yellow macaw named Charlie, or as she (parrot sex is not easy to determine) is also sometimes known, Charlie the Curser. Charlie's owner received international press attention in 2004 when he went public with the claim that Charlie had once belonged to Winston Churchill, the famous wartime prime minister of England, and that Churchill had taught her to say "[Bleep] the Nazis" in an excellent imitation of Sir Winston's voice. If true, Charlie, who is still alive as of this writing, would now be at least 116 years old. The only problem with the story is that no one who was in Churchill's household at the time remembers him having a macaw. No photos of Charlie with Sir Winston can be located, and no one except Charlie's owner has ever heard her swearing at the Nazis or anyone else. Churchill's daughter Mary says that the story is absolute nonsense. But the story has done wonders for business at the garden shop where Charlie lives, as people come from far and wide to see Churchill's parrot.

My wife and I own a parrot named Hector, a yellow-headed Amazon parrot that may be seventy years old—or not. We know we have had Hector for about thirty-five years and the people who gave her (yes, her, I told you sex is difficult to determine in parrots) to us said that she was thirty-five years old at the time. However, we never tried to get hard documentation of that age from the original owners (if they were the original owners). It would be easy enough for me to claim that I am the proud owner of a seventy-year-old parrot, but in fact, I just don't know. In the rest of this book, I will often mention how we know the real age of animals of special interest because, as physicist Richard Feynman once said, I'd rather not know something than think I know it and be wrong.

Before we set off on our journey, let's not forget about Dr. Dunnet's fulmar. After Dunnet died in 1995, I allowed some time for his friends to mourn and then contacted Paul Thompson at Aberdeen University, who had taken over the northern fulmar study. Was that bird—you know, the famous one with George as a young man and again as an old man—still alive? Yes, it was, I learned. It survived at least one year longer than George did before disappearing for good—likely lost at sea—at the age of at least forty-eight years young. Amazing.

The northern fulmar is a long-lived bird, but birds are long-lived as a group. So are bats, it turns out. Why? Could it be flight itself that conveys long life? Only four groups of animals have ever evolved the ability to fly. Let's see if they all evolved long life, too.

I LONGEVITY IN THE AIR

2 THE ORIGIN OF FLIGHT

If you have ever paddled or poled your way through a Louisiana bayou—a dark, flooded swamp forest—while sunrays slanted through the feathery cypress canopy and the still, black water felt particularly menacing, then you have a pretty good idea what it would look and feel like to float through the flooded Carboniferous forests of 380 million years ago. Unseen alligators and crocodiles provide a bit of menace in modern swamp forests. Man-size water scorpions and carnivorous, salamander-like amphibians were their counterparts lurking beneath the water of the Carboniferous swamps.

This ancient forest was noticeably silent. No birds chirped or sang, no frogs croaked or chorused, no mammals yelped or yowled. Evolution had not yet conjured these animals into existence. It was a visually as well as acoustically drab world. Flowers had not yet evolved to brighten the forest's greens, grays, and browns. However, in this drab, quiet world, one of the most momentous events in the history of life took place. Animals, for the first time, were taking to the air, something that later would have widespread consequences for life in general and for animal longevity in particular.

Plants took to the air first. Once they had blanketed the bare earth, they had nowhere to go but up—first knee, then waist, then chest high. Within the geological blink of an eye, they had shot up to towering thirty-meter (hundred-foot) tree fern forests. Competition drove the upward mobility of plants—competition for precious sunlight, plants' life-giving energy source.

Earth-hugging plants like mosses, horsetails, and liverworts first covered the ground, but after that, any evolutionary innovation that sent plants sprouting higher would have been competitively favored. Taller plants could intercept the sun's energy first, their shade in turn depriving their shorter neighbors of it. It was a race for light by height.

Where the plants went, animals quickly followed. They crept up stalks, sticks, and trunks, along branches and twigs, to find new food sources high above the ground. The trees also provided refuge from earth-bound enemies. These early arboreal animals pierced, sucked, sawed, rasped, and chewed the treetop greenery as the forest grew. Species numbers in the trees grew rapidly. The smallest arboreal animals dined on bacteria or fungi growing on the plants' surface. Larger predatory animals fed on them.

Animal life in these early forests was dominated by arthropods—millipedes, centipedes, scorpions, pseudoscorpions, whip scorpions, spiders, and a host of creeping, leaping, hopping, insects, the group that would dominate the world ever after. It would have been a horror gallery for those who shudder at crawlies, especially if some—such as millipedes—grew to the size of Komodo dragons. Insects emerged from the water soon after the first plants began colonizing the land. As terrestrial plant communities grew more complex in structure—a hodgepodge of ground mats, upright plants, shrubs, vines, and trees—insects also diversified.

BODY SIZE AND FIRST FLIGHT

The first insects to take to the sky were small. This is because gravity is forgiving to small things. Understanding why body size affects nearly every facet of biology including longevity will be central to understanding how long life evolved, so a brief explanation about how it influenced the origin of flight deserves a small digression.

Imagine that you are standing on the windy eighty-sixth floor open-air observation deck of New York City's Empire State Building. This is many times higher than the highest rainforest tree in the world. You have with you a bottle of tiny, wingless flies. You open the bottle and dump them over the building's side. What happens?

We all know this intuitively. They do not plummet to the ground as would a cannonball, a coin, or your car keys. Because they are both small and light, when they fall, they fall gently. In fact, a slight updraft might whisk them even higher into the air for a time and carry them for miles before eventually depositing on the ground.

The reason small wingless insects fall slowly or can be swept even higher by a slight breeze is that by virtue of being tiny and light, they have a large surface relative to their weight. Air pushes back against surfaces, so the resistance of air slows their fall. This is why kites fly and feathers float on windy days. Kites and feathers are designed in shape and materials to have extremely large surfaces relative to their weight so they can be pushed around easily by the air.

Gravity poses little danger to small animals because small size *ensures* a large surface relative to weight. Simple geometry tells us that an ice cube one centimeter on a side, as you might find in a fountain drink, has a hundred times as much surface area relative to its weight as a one-meter block of ice, even though they are the same shape and made of exactly the same material. So small size increases surface-to-weight ratio. Not only does this ratio affect the speed at which objects fall, but it also affects the way objects interact with their environments in other ways. For instance, because heat is transmitted across surfaces, the large surface area of a small ice cube relative to its volume means it will much more quickly melt in the heat and freeze in the cold than will a large block of ice. Similarly, small animals are more vulnerable and large animals are more resistant to temperature extremes.

Getting back to gravity, the converse also holds. In terms of falling, gravity becomes more important and air resistance less important as animals become larger. The practical consequence of this geometry has been vividly described by biologist J. B. S. Haldane, who wrote, "you can drop a mouse down a thousand-yard mine shaft, and, on arriving at the bottom, it gets a slight shock and walks away. A rat is killed. A man is broken. A horse splashes."[1] Haldane as a boy helped his father investigate mine accidents, so these descriptions may come from some firsthand knowledge.

BACK TO FIRST FLIGHT

So for those first tiny insects that reached the canopy of thirty-meter tree ferns in the Carboniferous forest, there would be relatively little danger from mistakes in leaping from twig to twig, limb to limb, even tree to tree. A fall would be unfortunate but not catastrophic.

The first stage of insect flight probably began exactly this way, leaping among the foliage in the thick canopy of these ancient forests. Individuals that could leap the farthest and alight the most accurately relative to their peers would be more successful at escaping predators, more efficient at quickly moving to previously unexploited food patches, and ultimately more successful at finding mates and leaving descendants than their less nimble relatives. You might think of this as the parachute stage in the evolution of flight, small size itself acting something like a fall-slowing parachute to allow the adventurous to float through the air some distance before landing.

To understand the next step toward the origin of flight and the ways this affected insect longevity, we need to consider the various trajectories of insect life.

Insects get their size and shape from their cuticle, the "shell" or exoskeleton that cloaks their bodies, giving them structural support. Unlike our own skeleton, which is inside our body, the insect skeleton covers the outside of the body something like a suit of armor. An exoskeleton has the advantage that it protects inner organs but the disadvantage that it limits growth. Adult Sir Lancelot could not wear the same suit of armor as child Sir Lancelot. Insects grow by shedding one cuticle after producing another from beneath it. During the short period when the old cuticle has been shed and new one is drying and hardening, their bodies can expand. That is, they can grow. The new cuticle might even have a different shape than the previous one, allowing insects to change shape as well as size as they grow. Thus, growth and ability to change shape allow them to adapt to different environments or to adapt differently to the same environment. The immature butterflies and moths that we call *caterpillars* look and behave very differently than their adult selves. Once they reach adulthood, when they are first capable of flight, insects cease molting. Insect adulthood abandons growth and change.

Back in the swamp forests of 380 million years ago, any chance mutations that lengthened or flattened any small outcroppings of the adult insect exoskeleton without sacrificing rigidity and strength would increase their surface area, allowing them to serve as airfoils so they could glide further than simply leaping. The identity of the specific exoskeletal outcroppings that ultimately lengthened, flattened, lightened, and later became movable wings has been debated for decades, because unlike any of the other three times that powered flight evolved in animals, insect flight did not repurpose an existing limb to serve as a wing. Wings were added to the basic six-legged insect body plan. You could think of the earliest phase as the parasail stage in the evolution of flight.

Additional chance mutations that lengthened or further flattened these cuticle outcroppings or that lightened them without compromising their stiffness, making them more effective airfoils, would have been favored by natural selection as would any mutations that made the outcroppings movable, so they could be tilted and turned for better maneuverability. Eventually—and we don't know when and in what insect group exactly because the fossil record is sparse during this period—these lengthened, lightened, and movable outcroppings evolved a complex joint where they met the body wall, and that joint acquired muscles to move it. Now all that was lacking to evolve powered flight—flight that could overcome gravity rather than just minimize its effects as sailing does—was the energy delivery kit to power wing movement.

THE WINGS' ENERGY-DELIVERY KIT

An energy-delivery kit is required to provide power to the wings. Powered or flapping flight, as contrasted with parachuting or gliding flight, is the most energetically demanding activity animals perform. However, a compensating advantage is that powered flight covers distance rapidly compared with other means of locomotion, so it is energetically *inexpensive* in terms of distance traveled.

How energetically costly is insect flight? Consider a human sprinting up a flight of stairs—among the most energetically expensive things humans can do. This activity may increase one's energy use as much as seven- to fourteen-fold relative to sitting in a chair reading. For comparison, insect flight, in which wings typically beat more than a hundred times *per second*—and in some species as fast as a thousand times per second—increases energy use fifty to 150 times compared to the same insect at rest. The energy-delivery system that powers flight consists of large, highly efficient flight muscles inside the exoskeleton to which fuel and oxygen must be delivered at exceptionally high rates. To put the wingbeat velocity of flying insects in context, a hovering hummingbird (hovering is the most taxing form of bird flight) beats its wings eighty times per second—or near the lower end of the insect wingbeat frequency range.

Insect flight muscles are huge, forming as much as 30 percent of an animal's body weight. These flight muscles are also aerobic, meaning that they are specialized for endurance rather than brief bursts of activity. Aerobic muscle, as the name, implies requires a continuous supply of oxygen.

At the level of the muscle cell, *aerobic* means that almost all energy is provided by mitochondria, the cellular organelles ("little organs") where oxygen breaks apart the chemical bonds of food to release energy. Diagrams of animal cells that we've all seen in biology textbooks are misleading when it comes to mitochondria. They show at most a handful of mitochondria per cell. This is necessary for illustrative purposes, but in reality, there can be hundreds or even thousands of mitochondria per cell, depending on the cell's energy needs. A developing human egg, for instance, contains half a billion mitochondria. Insect flight muscle cells are also packed with mitochondria, which comprise as much as 40 percent of their volume. In terms of aging and longevity, mitochondria are especially important in that as they produce energy, they simultaneously produce damaging free radicals. Free radicals are molecules that damage other biological molecules with which they interact. Consequently, the necessity to ramp up mitochondrial activity for flight was potentially accompanied by a massive increase in free-radical production by flight muscles. Damage to muscle cells from these radicals would degrade those cells over time and shorten the lives of flying insects relative to less energy-demanding nonflying species.

Energy delivery and use as well as the accompanying production of oxygen radicals are critical components of longevity. We will revisit them often in future chapters.

LONGEVITY OF FLYING INSECTS

Powered flight gave insects a tremendous advantage in colonizing the earth. In terms of species numbers and, in fact, in terms of total numbers, they far exceed that of any other animal group, even today. By some estimates, there are as many as 400 million insects per acre of land on earth. More strikingly, perhaps, entomologists recently calculated that there are about 17 million flies—just flies—on earth for every human. The total weight of all flies on earth is greater than the total weight of all humans, in fact—and flies are only one of about thirty major groups of insects.

Among birds and mammals, as we shall see, the ability to fly is intimately associated with exceptional longevity. Do flying insects, such as flies, bees, wasps, butterflies, dragonflies (indeed, most insects), live longer than non-flying ones, such as silverfish, bristletails, and others in which wings never evolved, or even those such as fleas and many stick insects, whose ancestors had wings but who lost them over evolutionary time?

The short answer is no. Flight in insects is not associated with exceptional longevity for reasons that reveal a lot about the requirements for exceptional longevity. In essence, this may be because insects aren't flying creatures at all. Instead, they are creatures that can fly.

What do I mean by that?

The mayfly illustrates this point nicely. Mayflies, the darlings of fly fishermen everywhere, appear in enormous numbers over streams in the spring or fall, but within a few days, they disappear. Historically, mayflies are animal poster children for brief lives. Mayfly "short-gevity" was noted by Aristotle. The Roman encyclopedist Pliny the Elder in his *Natural History* claims that mayflies do not live beyond one day, and for a number of species, this is true—in a limited sense. Even the name of the insect order to which mayflies belong—*Ephemeroptera*—suggests short life.

But in fact, mayflies do not have particularly short lives for something as small and vulnerable as an insect. Mayflies, like any number of other insects such as dragonflies and stoneflies, spend their preadult lives as wingless aquatic nymphs living in streams, molting through a series of successively larger nymphal stages, before finally emerging as flying adults.[2] So here is a key fact about mayflies. Their juvenile life in streams lasts several years. It is only their adult life that is short. A life of several years is not short for an insect. It is long, in fact. We have been misled about the longevity of many insects because traditionally, we have focused only on adult life.

Insects also differ from all other flying creatures in that they gain the ability to fly only as adults. This is what I meant when I said that insects are not flying creatures at all but creatures that can fly. For some insects, wings do not even last all of adulthood. Ant queens, for instance, have wings for a brief period after becoming adults, but once they have completed their "nuptial flight," during which they mate, they purposely break them off before digging into the earth to start their own colonies.

All juvenile insects are earth- or water-bound. So flight does not define and therefore is not critical to insect survival for much of their lives. Many insects, unlike animals such as ourselves, spend most of their lives as juveniles. The adults may have wings, but wings hardly contribute to their longevity because the majority of their lives are spent wingless, buried in the soil, crawling on leaves, or hiding among pebbles in streams.

Periodic cicadas may be perhaps the most famous example of insect longevity. Periodic cicadas emerge synchronously in eastern North America in stupefying numbers every few years. Males clinging to tree trunks making a deafening droning buzz that can be louder than a passing train. The buzz thankfully lasts only a few weeks because adult cicadas live only a few weeks. However, before emerging *en masse*, they spent as much as seventeen years living underground as immature nymphs, feeding on root sap. Taken as a whole, their life of almost two decades is among the longest of any insect. It is only their life when they are visible to us that is short.

So if an animal lives a long life as a juvenile and only a short one as an adult, what does that say about the rate at which it ages? Aging and longevity are closely related but not identical concepts. People are at least as concerned about *how* they will age—meaning how they will deteriorate physically and mentally as they grow older—as they are about *how long* they will live. No one wants a long life of frailty and decrepitude like Tithonus, the man in Greek mythology who thanks to a curse from Zeus could not die but grew increasingly frailer and weaker as time passed.

Think of it this way. The life cycle of most animals consists of a period of development—growth and change—followed by adulthood. Development is often accompanied by *improving* physical function with age, the opposite of aging. Think of a ten-year-old child compared to her earlier two-year-old self. At ten, she is physically superior in virtually every aspect, including being better able to survive, than she was at age two. Adulthood is the life stage typically associated with reproduction and deteriorating physical function or, as I use the term here, aging.

In insects, as soon as molting ceases, aging or deterioration of the adult body begins. In some insect species such as mayflies, this deterioration is phenomenally rapid. In others, it occurs gradually over years or decades.

In insects, these phases of life are particularly separable because of molting. Insects may go through many molts during their immature life, but once they reach adulthood, molting forever ceases.

Most reports of insect longevity focus only on adults. For our purposes, though, we want to understand the entire lives of animals from egg to grave. And from this perspective, there seems to be no discernible pattern of greater longevity for flying insects compared to nonflying insects. Among the tremendous diversity of insect life histories, there are both short- and long-lived flying species, but most of the differences are in how long it takes them to reach adulthood rather than in how long adults survive.

Among actively flying insect species such as dragonflies, butterflies, or actual flies, none are particularly long-lived *as adults*. Virtually all live only weeks or at most a few months (figure 2.1). Probably none live more than about a year in nature. In one of the largest field studies focusing on adult insect longevity, led by Jim Carey of the University of California, Davis, for instance, researchers marked more than thirty thousand butterflies over a nearly four-year period in Kibale Forest National Park in western Uganda.[3] If you expect to find long-lived insect adults anywhere, it would be in the tropics where there is no cold, food-deprived winter. The researchers noted each time they spotted one of the marked individuals again. For all this effort, the longest time between marking and last sighting of any individual was about ten months for a widespread forester (*Euphaedra medon*) butterfly. They also collected over six hundred other butterflies and kept them in large outdoor cages where they were protected from predators. None of these survived longer than about three months. I've inquired of many field entomologists, and, with several striking exceptions to be described in a later chapter, none knew of any species of insect that had been documented to live more than about one year as an adult in the wild.

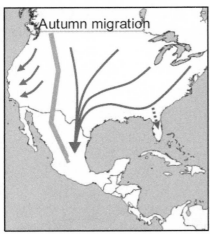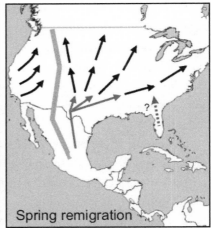

Figure 2.1

Longevity and the Monarch butterfly (*Danaus plexippus*) migration. Virtually no insects perform regular two-way migrations as do many bird species. Monarchs of the eastern United
States are an exception—an exception that illustrates an intriguing longevity pattern. After
wintering in colonies of millions in central Mexico, they begin a multigenerational northward
migration. The overwintering generation lay eggs in Texas and Oklahoma and then dies. The
new generation continues north, living a couple of months before laying eggs and dying, hand-
ing the baton, so to speak, to yet another short-lived generation. This generation continues
north, lays eggs, and like its parents dies after a short couple of months. But these eggs are
different. Whereas the other generations live only about two months as adults, this generation
will survive as long as eight months, fly several thousand miles back to Mexico, live through the
winter, and then head north again in the spring. A key to this remarkably longevous generation
is that they hormonally shut down their reproductive activity for the first six months or so of their
adult lives. It is reproduction that apparently limits their longevity. The presence or absence of a
substance called *juvenile hormone* is what turns reproduction and aging on and off.

Sources: W. S. Herman and M. Tatar, "Juvenile Hormone Regulation of Longevity in the Migratory Monarch Butterfly," *Proceedings of the Royal Society B: Biological Sciences* 268, no. 1485
(2001): 2509–2514. Figure modified with permission from *Journal of Experimental Biology*.

If flight—even of just adults—is associated with longevity in birds and
bats but is not in the insects, why not?

The answer may be that while an exquisite flight kit evolved in insects,
a particularly good longevity kit did not. A key component of a successful longevity kit is the ability to repair or replace damaged parts. Actively
flying insects (indeed, any insect) cannot repair or replace damaged wings,

mouthparts, or legs as adults. Once they have finished molting, insects have an extremely limited ability to repair their cuticle, which includes their wings. Unlike vertebrate bone, which is a living tissue, the insect cuticle (the outer layer of the exoskeleton) is composed of nonliving material—more like fingernails than like skin. When nonliving material is damaged, it stays damaged, as anyone who has damaged a fingernail can attest. Immature insects can repair damaged parts during molting because as the old cuticle is shed, replacement parts and pieces are able to sprout before the new cuticle hardens. So mayflies swimming in streams or cicadas feeding on roots can molt and eventually replace a lost leg or antenna they may have suffered. However, as wings do not appear until adulthood when molting is finished, wings are not reparable. Wing damage likely limits the lives of insects in nature.

And wing damage is unavoidable. Think of the simple wear and tear that can occur to wings pounding through the air at a hundred to a thousand times per second any time an animal flies. If you head outside with an insect net and capture almost any flying insect with visible wings, you will notice that almost any individual you capture will have nicks, chips, or cracks in its wings. Notice that I said species with "visible wings," by which I mean things like butterflies, dragonflies, and house flies, where wings are continuously exposed. In most species of beetles and some of the true bugs, the delicate flight wings are protected when the animal is not flying under thicker forewings called *elytra*, so examining the flight wings is not easy. Damaged wings compromise flight, making the animals vulnerable to predators and less able to forage widely. So it may not be surprising that although adults of flying insects are not long-lived, most surviving in nature no more than weeks to months, the longest-lived appear to be beetles with their well-protected flight wings and their habit of flying relatively little. This vulnerability to flight damage also likely limits the evolutionary advantage of developing particularly efficient internal defenses against damaging cellular processes such as the production of oxygen radicals.

So even though flight is often associated with exceptional longevity, this isn't true among the insects. The longest-lived insects will not be found in the air but elsewhere.

3 PTEROSAURS: THE FIRST FLYING VERTEBRATES

For more than 100 million years, insects had the skies to themselves, giving them an unparalleled opportunity to spread far and wide, diversify, multiply, and form thousands upon thousands of new species. This initial advantage may be why they became by almost any measure the largest and most successful animal group on earth. Yet even insects during this time of expanding opportunities never completely emancipated themselves from water.

Water is life. The chemical reactions that power life take place in the water inside us. The search for life on other planets begins with a search for water. Desiccation was the great danger that animals (and plants) faced as they moved onto land. Insects and other arthropods such as spiders, scorpions, and millipedes had their solid cuticle to help prevent water loss from their internal tissues, but their eggs were small (large surface relative to their volume) and thus particularly vulnerable to evaporative desiccation. Eggs needed to be placed where it was moist—in or near water. Yet a small, unobtrusive, lizard-like creature barely visible among the giant insects, millipedes, and amphibians found a way to abandon the water entirely as its eggs became surrounded by a leathery, evaporation-resistant but breathable shell. As the Carboniferous swamp forests collapsed, leaving cooler, drier, more open forests behind, reptiles—with their evolutionarily innovative, desiccation-resistant egg and their scaly, waterproof body covering—found their own niches on land.

Reptile dominance awaited a cataclysm though. That cataclysm—the end-Permian extinction some 250 million years ago—came as close as any known event to extinguishing life on earth. More than 95 percent of marine species and 70 percent of all terrestrial species disappeared, as did more than half of all insect families. The cause or, more likely, causes of the Great Dying, as it is sometimes known, remain obscure. Massive volcanic activity was involved, possibly a collision with a large extraterrestrial body (an asteroid or a comet), or perhaps even a chemical catastrophe that turned the oceans and the atmosphere temporarily poisonous. It may have even been some combination of these. What is known, however, is that life recovered, and when it did, reptiles emerged to dominate the land.

Before long, powered flight had evolved within reptiles for the second time in the history of life. That group of flying reptiles is called *pterosaurs* (figure 3.1). The complete fossilized skeleton of a crow-size flying reptile, colloquially called a *pterodactyl*, was discovered in the late eighteenth century in the famous Solnhofen limestone deposits of southern Germany. This well-preserved pterosaur skeleton embedded in the limestone was so

Figure 3.1
Pterosaur size and flight. All pterosaurs, even the largest species, apparently could fly. Among the largest were *Hatzegopteryx* (middle) and *Aramourgiania* (right). The giraffe and human are shown for scale.
Source: Drawing courtesy of Mark Witton.

different from any living animal or previously known fossil that it was not properly identified by the French anatomist, George Cuvier, as a reptile with wings until several decades later.

Today pterosaurs are represented by a remarkably rich and diverse fossil record. More than one hundred species have been identified, which means that there were many more in their heyday—the age of the dinosaurs. Some pterosaurs were as small as blackbirds, and others as tall as giraffes with the wingspan of small airplanes. They were also widespread. Pterosaur remains have been discovered on every continent except Antarctica, and there is little doubt they lived there as well, but their fossils are buried under kilometers of ice. In addition to fossilized bones, paleontologists have found well-preserved three-dimensional pterosaur specimens, extensive soft-tissue remains, excellent impressions of skin, and well-preserved eggs with well-preserved embryos inside them. They have even identified foot tracks of pterosaurs, verifying that they used their winged forelimbs for walking and running as well as flying. Pterosaurs were indeed unlike anything seen before or since, except perhaps in the popular television series *Game of Thrones*, where the form and function of fictional dragons was shamelessly appropriated from pterosaurs.

Although often called *flying dinosaurs*, pterosaurs are not dinosaurs. They were a sister group to dinosaurs. Their wings stretched from an absurdly elongated fourth finger (yes, it is called the wing finger) to their ankles. Their wings and the associated shoulder bones and muscles (called the *shoulder girdle*) were dual use—so far as we can tell, reasonably well-designed for flight in all species while retaining their use for locomotion on the ground or clambering about in the trees. Some even apparently swam. Unlike the modern flying vertebrates—birds and bats—pterosaurs were truly quadrupedal. The only living species with somewhat similar capacities are a few bats, most famously the vampire bat, which can walk—and even run—on two legs and two "palms" as well as fly.

VERTEBRATE INVENTIONS

Pterosaurs were vertebrates, meaning that, like us, they had backbones. Vertebrates as a group differ biologically in some important ways from insects.

Most obviously, they—that is, we—have an internal skeleton rather than an external skeleton like insects. This allows us to grow continuously rather than in spurts associated with molting as the insect exoskeleton dictates. Continuous growth requires the continuous production of new cells by division of existing cells to form new tissue. Insects have very limited continuing cell division as adults. Only a few cells in their digestive tract and reproductive organs maintain the ability to make new cells. As a consequence, they have almost no ability to repair damaged body parts, as noted earlier. Because an internal skeleton does not protect the body's external surfaces as a hard exoskeleton would, continuous cell division in vertebrates is necessary to repair external wounds of any sort. The vertebrate adaptive immune system, in which specialized blood cells recognize, attack, and *remember* foreign invaders, also require the ability to pump up cell division to fight off invaders. However, retaining the capacity for cell division throughout life comes with a price. As long as cells are capable of dividing, that division can potentially lurch out of control. Uncontrolled cell division is cancer—a problem for vertebrate longevity with which insects don't have to contend.

Pterosaurs were unusual reptiles in numerous ways in addition to their ability to fly. They were furry, for instance. Their short fur, which covered most of their body and head except for jaws and wings, apparently performed the same function as it does in mammals—insulation—suggesting that they were likely warm-blooded. *Warm-blooded* is actually a rather misleading term. What is typically meant by it is *endothermic* (inside heat). That is, some animal groups like birds and mammals—and apparently pterosaurs—produce their own body heat from the metabolic activity of their internal organs rather than having to rely on external sources such as the sun to warm them. The great advantage of endothermy is that by maintaining muscles and other organs at peak operating temperature all the time, animals can remain highly active at any time of the day or year regardless of the ambient temperature. The great cost of endothermy is that continuously producing your own body heat is energetically expensive. If pterosaurs were endothermic, then their metabolism—the rate at which they used energy and the amount they would have to eat to provide fuel for that energy—would have been considerably higher than an equivalent-size cold-blooded, or more properly *ectothermic* (outside heat), reptiles like those we have today. The cost of endothermy requires a human to eat about twenty-five times as much food per year as a human-size alligator.[1]

Because generating internal heat is energetically costly, conserving that heat to the extent possible—rather than continually losing it to the environment—makes sense. Insulating the body surface with fat or fur or feathers reduces heat loss in the same way that an insulated cup keeps your coffee warm longer. Heat loss would have been a particular problem for small pterosaurs because of their large surface (across which heat is lost) relative to their weight (heat-producing tissue). Because they were endothermic, at least to a degree, pterosaurs could not only fly, but they could remain active at times of the year or times of the day—such as early morning or night, when their ectothermic cousins would have moved sluggishly at best, making them easy pterosaur prey.

The diversity of pterosaur forms, although constrained by the demands of flight, was impressive. Some species had long bony tails, enormous colorful head crests, and typical reptilian teeth. Others lacked tails or crests and developed bird-like toothless beaks. Some had long necks and dagger-like beaks, like the herons or cranes of today.

THE ROUTE TO REPTILIAN FLIGHT

The details of the evolution of pterosaur flight are poorly documented as the fossil record during the time flight first arose is sparse. However, most plausibly, it followed the insect pattern. In one sense, it would have been easier for reptilian flight to evolve because it required only the modification of existing forelimbs rather than the *de novo* invention of completely new wing structures as in insects.

The first stage of reptile flight evolution would no doubt involve small reptiles that lived by scampering and leaping about in trees, relatively safe from bungled landings because of their size, evolving powerful forelimbs for climbing and catching themselves when leaping among branches. Instead of their legs being splayed to the side like modern lizards, their legs would have been tucked beneath them, thereby enhancing their jumping ability. Standing on a level surface, they would have had an erect posture more like a squirrel than a lizard.

Any fortuitous genetic mutations that would have lengthened the skin between forelimb and hind limb would have turned these leapers into potential gliders, those skin flaps serving as airfoils. Gliding apparently evolves easily. As an effective energy-conserving method of moving among trees, it has appeared again and again. Some of the earliest mammals, contemporaries of the pterosaurs, were gliders. Gliding in today's animals is probably more common than you realize. There are gliding squid, gliding frogs, numerous gliding lizards, gliding snakes, and more than fifty species of "flying" (that is, gliding) fish. Among mammals, there are dozens of species of "flying squirrels," three groups of gliding marsupials (the best-known of which is the sugar glider), and two species of colugos (sometimes called "flying lemurs," even though they glide, not fly, and are not lemurs).

But the evolution of powered flight is clearly much more difficult than the evolution of gliding, or it would have evolved more than four times in the past 700 million years. Even small proto–flying reptiles would have had to deal with the weight problem because even small reptiles are many times the weight of flying insects.

One way that evolution dealt with the pterosaur weight problem was by reducing the skeleton. Pterosaurs did that more and more over the course of their 150 million–year history. The bony reptilian tail typically found in early pterosaurs became vastly reduced in later evolving species. Later species also had fewer teeth, and the heavy reptilian jaw was replaced by a much lighter, bird-like beak. Pterosaurs also gradually evolved more and more hollow bones, strengthened internally by struts. In early pterosaurs, only the skull and some parts of the spine were hollow, but in later species, almost all of the bones associated with the wings (the shoulder bones) were hollow as well. This may have been for weight reduction. However, it might have also served another purpose—enhancing oxygen delivery to the flight muscles. Hollow bird bones are part of their unique, highly efficient breathing apparatus. They contain air sacs connected to the lungs. Certainly, a highly efficient respiratory system for delivering enough oxygen to demanding flight muscles is a necessary component of any flight kit. The rest of that kit was highly mobile forelimb joints so that those limbs could be used in flapping flight as well as in walking and running on the ground.

The final piece of the kit is the massive musculature and associated sturdy bones to which those muscles could be attached to flap the wings. This powerful musculature was probably also critical for takeoff from the ground. Vampire bats are the model for this idea. They become airborne with a quick powerful launch using all four limbs like an explosive pushup. As we know, pterosaur forelimbs were designed for locomotion on the ground as well as in the air, so it makes sense that they would have used the power of those forelimbs to help launch themselves into the air.

As with actively flying insects, the massive flight muscles of pterosaurs would have been packed with mitochondria pouring out the energy needed for flight but also pouring out damaging free radicals. To maintain the power and efficiency of these muscles year after year, they would have had to develop effective antioxidants, chemicals that detoxify free radicals, to minimize damage to the flight muscles. Or else they would have had to evolve a highly efficient ability to repair muscle damage, possibly using muscle stem cells as humans and other mammals do. If they failed to develop these defenses in key flight muscles, they would have been unlikely to live very long, bringing up the key question of how long pterosaurs lived. Were they long-lived enough to make them extinct members of Methuselah's Zoo?

HOW LONG DID PTEROSAURS AND DINOSAURS LIVE?

To address the longevity of pterosaurs compared with, say, the earthbound dinosaurs of the time, we need to match them for size. Like virtually every other animal group, we would expect large dinosaurs and large pterosaurs to be longer-lived than smaller ones. A number of dinosaurs were much larger than any pterosaur, of course, but there were plenty of species of both that were similar in size. The smallest dinosaurs weighed about two kilograms (4.4 pounds).

You might be surprised to learn that we can actually make pretty good guesses about how long at least a few dinosaurs lived, given that they have all been dead for at least 66 million years. We can do this because a number of species' bones contain growth rings, something like trees. Yes, some animals continue to grow as adults, and for those that do, their bones often show growth rings because seasonal variations in temperature and food availability will affect bone growth rate. Examining the bones of modern crocodiles of known age, for instance, we have learned that *their* growth rings are produced annually. Each ring represents a year of life.

Assuming that the same thing held true in dinosaurs, we can estimate how fast they grew (by the distance between adjacent growth rings) and how long they lived. For instance, *Tyrannosaurus rex*, which was roughly the weight of a male elephant, we know reached maturity in about eighteen years, which is a bit slower than elephants reach maturity. However, dinosaurs seem to be considerably shorter-lived than you might expect. According to its bone growth rings, the oldest known fossil *T. rex* lived only twenty-eight years.[2] Keeping in mind that we have only a few adult *T. rex* specimens to examine, it is likely that some lived considerably longer—perhaps even several times longer. However, from what we know now, a twenty-eight-year maximum is less than half as long as an average mammal of that size is expected to live. A much smaller eighteen-kilogram (forty-pound) plant-eating dinosaur called *Psittacosaurus mongoliensis*, a relative of *Tricerotops*, lived only ten to eleven years based on the small number of available fossils. The longest-lived dinosaur known so far was a fifty-four-foot-long, twenty-ton behemoth with a name, *Lapparentosaurus madagascariensis*, nearly as long as it was, that survived only into its early forties—again, so far as we know from a few individuals,[3] about half as long as we would expect of a mammal that size. So dinosaurs, at least from the limited evidence we have (emphasis on *limited*), seem to have been relatively short-lived as a group. For comparison, elephants in nature routinely live into their fifties, less frequently into their sixties, and occasionally into their seventies.

Disappointingly, we know next to nothing about the longevity of pterosaurs. Their bones were hollowed out so thoroughly to reduce weight for flight, that growth rings once they reached adult size are simply not available. Given their resemblance to birds in many aspects of their anatomy, we can speculate that they would have been considerably longer-lived than dinosaurs. All we know for certain are the growth rate and the age at maturity in a few species. Growth has been measured best for a pterosaur named *Pterodaustro guinazui*, which was extraordinarily odd-looking even for a pterosaur. With a ten-foot wingspan—a bit larger than the wingspan of a stork—it had a long neck and long, upswept jaws with as many as a thousand very fine teeth that resembled the baleen in a filter-feeding whale's mouth. So it probably also filter-fed from the sea a bit like modern flamingos feed from lakes. We estimate from its growth rate that it probably reached sexual maturity at about two years of age and reached its maximum size at about six years, although we don't know how much longer than that they lived. Although these ages don't suggest that pterosaurs were particularly long-lived, they don't rule it out either. One observation suggesting that pterosaurs may have been long-lived is that surprisingly often, we can see that they had arthritis in their wing joints. Arthritis is a result of the wear and tear of aging, so pterosaurs lived at least long enough to become arthritic.

Having evolved on the heels of a global cataclysm, the end-Permian extinction, pterosaurs went out in a cataclysm, too. Sixty-six million years ago, a massive asteroid slammed into the earth near the Yucatan peninsula, initiating a series of events that exterminated three-fourths of all species on earth, including all the "traditional" dinosaurs and all the pterosaurs. However, pterosaurs may have already been on their way out. We know of only a few, large pterosaur species that were still around when the asteroid hit. The smaller species seem to have already vanished. No one is exactly sure why, but they may have suffered from competition from some other flying creatures that by this time were thriving. Paleontologists call these creatures *theropod dinosaurs* with wings and feathers. The rest of us call them *birds*.

4 BIRDS: THE LONGEST-LIVED DINOSAURS

Birds are masters of the air. They fly higher, faster, farther, dip, flip, and dive with more agility, and remain aloft longer than any other group of flying animals. The agility of many predatory birds is such that it allows them to routinely snatch other flying animals—insects, bats, and other birds—directly out of the air. In fact, the presence of birds may be what forced vast numbers of other flying species into the night when few birds are active.

Seeing birds on a daily basis, it is easy to forget how remarkable their physical feats are. The bar-headed goose, for instance—without training, without acclimatization, without supplemental oxygen—takes off from sea level, climbs to nine thousand meters (thirty thousand feet) (where the temperature can be dozens of degrees below zero), and flies over the Himalayas (the world's highest mountain range) in less than twenty-four hours. A wading bird called the bar-tailed godwit flies nonstop for nine straight days and nights without refueling (that is, without alighting to eat or drink) on a eleven-thousand-kilometer (seven-thousand-mile) migration from Alaska to New Zealand. That, by the way, is within shouting distance of the world's longest nonstop commercial airline flight. A common swift has stayed aloft continuously for more than ten months. A racing pigeon can fly 160 kilometers (a hundred miles) per hour horizontally, and a peregrine falcon dives on its prey at up to 320 kilometers (two hundred miles) per hour. Even common garden birds perform spectacular, if largely unappreciated, acrobatics. Think, for instance, how often you see a bird abruptly drop from its cruising speed of forty kilometers (twenty-five miles) per hour to land precisely and with perfect balance on a pencil-thin twig.

But perhaps the most striking physical talent of all is the avian knack for longevity.

One illuminating example is the house sparrow (*Passer domesticus*)—that small, drab, black-masked and black-bibbed bird you are likely to see flitting about in back yards, parks, and gardens around the world. To highlight birds' remarkable longevity, let's compare it to the house mouse, a mammal of similar size with an eerily similar history. Both species originated in the Middle East. Both learned early on that grain planted and stored by humans was a more reliable food source than anything nature offered and so decided it was worth hanging close to people to exploit. Both spread across the world with the European diaspora and are now found on every continent except Antarctica.

But in terms of longevity, there is no comparison. Recall that a house mouse in the wild lives three to four months on average and a little more than a year for the longest survivors. In the laboratory, with all the care and coddling we can provide, the oldest ones manage to make it to about three years of age. The longest-lived *wild* house sparrow, on the other hand, has survived at least nineteen years and nine months,[1] or nearly twenty times as long as the longest-lived wild house mouse and more than six times as long as a mouse living a cosseted life in the laboratory or your home.

What longevity secrets allow birds to do it?

BIRD ORIGINS

Birds arose roughly 150 million years ago during the heyday of the pterosaurs. How were birds able to compete with and ultimately replace the pterosaurs that had at least a 50-million-year head start on them in adopting and refining an aerial lifestyle? Perhaps it had something to do with the fact that birds walk and run on two legs rather than four and use their forelimbs exclusively for flight rather than for both flight and terrestrial locomotion as did the pterosaurs. With no need for the compromises in form and function to serve both purposes, bipedalism may have allowed the rapid development of superior wing design. Wing design or perhaps some other trait may have allowed birds to perform superior aerobatics or be active at times of the day or seasons of the year when pterosaurs could not. A major contributor to the birds' success may have been the evolution of feathers, nature's crowning achievement in combining light weight with effective insulation. Pterosaurs, remember, were covered by a type of short fur. For whatever reason, despite the existence of dozens to hundreds of species of pterosaurs already filling the air, bird evolution rapidly took off.

The heyday of the pterosaurs when the first birds appeared was also the heyday of dinosaurs. This should not be too surprising as modern biology places the birds *among* the dinosaurs. Paleontologists these days commonly call what we call dinosaurs *nonavian dinosaurs* to distinguish them from *avian dinosaurs*—that is, birds. This deepens the mystery of bird longevity because as we saw in the last chapter, dinosaurs—at least from the few species for which we have evidence—appear to have been relatively short-lived. Remember twenty-eight years is the oldest *T. rex* found so far, which for an animal the size of an elephant is very short-lived, indeed. So at some point in the transition from terrestrial scaly reptile to feathered flier, the avian toolkit for long life evolved.

The first ancient bird fossil was discovered in 1861 in the same smooth limestone deposits in southern Germany where the first pterosaur fossil was found. *Archaeopteryx* (ancient wing) was a crow-size animal, the spitting image of a small, carnivorous dinosaur. In fact, it looked like a miniature version of *Velociraptor*, the agile and eerily clever predator made famous by the *Jurassic Park* movies. It had dinosaur teeth, a long bony dinosaur tail, *Velociraptor*-like claws on its forelimbs, and a large scimitar of a killing claw on its hind feet. It also had wings and feathers.

Actually, it isn't completely clear that *Archaeopteryx* should be called a bird. It is a classic transitional species caught between one form and another. It is likely, although not certain, that *Archaeopteryx* could fly. Although its wing feathers closely resemble the flight feathers of modern birds, it lacks the sturdy shoulder girdle and solid, keeled breastbone of modern flying species. Also, its shoulder joint doesn't appear to allow sufficient range of motion for its wings to flap high up over its back. It may have been able to flap enough only to give it an extra boost when springing into trees to catch prey or escape becoming someone else's prey, like modern grouse or turkeys. If it could fly at all, it wasn't a long, strong flier.

But before long in geological time, bird species with all types of flight abounded. As the smaller pterosaurs disappeared, more bird species appeared. Whether this pattern was due to direct competition between the birds and pterosaurs isn't clear. Birds on the whole were smaller than pterosaurs, the largest of these early birds being about the size of a goose.

The massive asteroid strike that finished off all of the nonavian dinosaurs and pterosaurs some 66 million years ago—in fact, that finished off three-fourths of all animal species on the planet—almost finished off the birds as well. Just a few species—the ancestors of the ducks and chickens and possibly the flying ancestors of the ostriches—survived. It was in the recovery from this near extinction that the ten thousand species of modern birds arose.

HOW LONG CAN BIRDS LIVE?

The exceptional longevity of birds has been appreciated for centuries. English philosopher and scientist Francis Bacon, writing in 1638, observed that birds generally live longer than mammals,[2] even though at that time he had only the vaguest idea of how long either wild birds or wild mammals actually lived. He knew the lifespans of only a handful of domesticated species, either farm animals such as chickens, sheep, and goats and pets such as dogs, cats, and parrots. Yes, the pet trade in parrots had already been thriving for centuries even in Bacon's time. About parrot longevity, he wrote that they have "certainly been known to live sixty years in England, in addition to [their] age when brought over." That would still be a reasonably accurate statement of what we know about pet parrot longevity today. He also knew that not all bird species were long-lived. The rooster, for instance, he noted "is lascivious, pugnacious, and short-lived." Yes, chickens, whether roosters or hens, are short-lived for birds. Maybe, as Bacon suggests, it is their moral shortcomings that make them so. Maybe it is something else.

When it came to wild, rather than farm, animals, Bacon simply made wild—and wildly inflated—guesses. Vultures, ravens, and swans all live about a century, he wrote, and elephants live two hundred years.

Absence of hard knowledge only enhanced birds' reputation for long life, though. Among pet birds today, longevity exaggeration is rife. Published but suspect (at best) anecdotes claim, for instance, that pet sulfur-crested cockatoos have lived up to 142 years or that Amazon parrots like my own pet Hector have lived 117 years. In fact, the firmly documented record longevity for these two species—records backed up by a birth certificate, so to speak—is a still-impressive fifty-seven years and fifty-six years, respectively. I guess that means either that my bird Hector has broken the world longevity record for her species by more than a decade or, more likely, that my wife and I were misinformed of her age when we acquired her.

The oldest captive parrot with firm age verification was a male Major Mitchell's cockatoo named Cookie, who lived most of his life at Chicago's Brookfield Zoo (figure 4.1). The name "Cookie," I suspect, is an American corruption of "Cocky," Australian slang for cockatoos. Major Mitchell's

Figure 4.1
Cookie, a Major Mitchell's cockatoo, the world's longest-lived bird, still looking fit and alert at age eighty-one. Birds are famous for maintaining health until almost the end of their lives. Cookie lived to age eighty-three, and in his later years, Chicago's Brookfield Zoo threw him a birthday party each year.
Source: Photo courtesy of Chicago Zoological Society/Brookfield Zoo.

cockatoo (*Lophochroa leadbeateri*, aka Leadbeater's cockatoo or the pink cockatoo) has strikingly salmon pink plumage from below, white plumage from above, with a bright red and yellow crest. It normally roams wooded parts of Australia's arid interior. Cookie arrived from Australia when the Brookfield first opened in 1934. He was a year old at the time. A popular zoo exhibit for decades, he was retired from public viewing in 2009 because of ill health. Cookie finally passed away to great sorrow and publicity in 2016 at the grand old age of eighty-three years. Putting Cookie's longevity in a mammalian context, if he were an average zoo mammal of the same three-hundred-gram (eleven-ounce) size, he would be expected to live no more than nine years.

What about birds in the wild? Is the house sparrow an exception, or do wild birds also live long amid the challenges of life of the natural world?

By the 1970s, birds' exceptional longevity was so well established that they had achieved a reputation for somehow avoiding aging altogether. Like Dr. Dunnet's fulmar, professional ornithologists might catch the same bird multiple times over several decades, and they seemed little the worse for wear as they grew older. Indeed, some researchers came to believe that their lives were limited only by unpredictable environmental hazards like storms, drought, or predator ambush, not by age. Using this logic, knowing the annual death rate of royal albatrosses from breeding colonies in New Zealand and assuming that their death rate does not increase with age, Yale University biologists Daniel Botkin and Richard S. Miller calculated that if a colony contained at least ten thousand birds (not unrealistic by albatross colony size standards) and if their annual death rate did not increase as they grew older, then at least one of the albatrosses alive when Captain James Cook arrived in New Zealand for the first time in 1769 should still be alive today, more than 250 years later.[3] Could wild birds really live this long? If they really remain healthy to the end of their lives, wouldn't it be nice if people could figure out how they manage it and achieve something similar?

HOW DO WE KNOW HOW LONG BIRDS LIVE?

We know far more about the longevity of birds in nature than any other group of animals thanks to Danish ornithologist Hans Christian Mortensen.

Around 1900, Mortensen invented a system for individually identifying wild birds by capturing them in live traps of various sorts and placing small, aluminum rings on their legs before releasing them.

You might think that Mortensen's invention would have made documenting bird longevity quite easy. It certainly made it much easier, but it still requires effort and persistence. To learn how long birds live, even if you can capture, recapture, and identify individuals, you still need to know their age when you first ringed them as well as when they die. Not all birds are like Dunnet's fulmars, tame and reliable about returning to the same nest site year after year, so that when they fail to return, you can be pretty certain they have died. Also, as the fulmars showed, documenting the longevity of some species may require studies lasting longer than any individual scientist's professional lifetime.

Still, thousands of bird enthusiasts and wildlife biologists have now been placing individually identifiable rings on birds for more than a century. Millions of wild birds from hundreds of species have now been ringed and recaptured multiple times. Just in North America, more than 64 million birds have been ringed since 1960. The British and Europeans have done the same. Centralized record keeping of all of this information from several continents now exists and is available at the touch of a mouse—the computer kind, not the laboratory kind.

I myself have placed rings on perhaps as many as a thousand birds, mostly in South America, and I must confess that there is something thrilling about briefly holding a living wild bird in your hands and watching it flap away in relief after you have placed that ring on its leg. I once caught an American golden plover (*Pluvialis dominica*), a species renowned for its long-distance migrations, in a mist net in central Venezuela that bore a numbered leg ring that I recognized as North American in origin. Sending the number off to the US Geological Survey's Bird Banding Lab in Maryland, I learned that the bird had got that ring two thousand miles away and four years earlier in Massachusetts. I had apparently intercepted it on its migration from its summer home in Patagonia back to its Arctic breeding ground. Amazingly enough, even though at least hundreds of thousands of these plovers migrate through the central Venezuelan savanna where I was working each year and I managed to catch only a few dozen of them, I netted the same bird in the same spot again a year later. So I learned firsthand that American golden plovers, despite migrating from southern South America to the Arctic and back again each year, lived at least five years. In actuality, we know from more than a thousand ringed American golden plovers that they can survive at least thirteen years of these arduous biannual journeys.

The hope is that if you catch enough individual birds of a species enough times, you will eventually have a pretty good idea about how long they live.

One thing we now know. No royal albatrosses live anywhere close to 250 years, as Botkin and Miller speculated. Birds, like nearly everything else, age. However, they do so slowly—so slowly, in fact, that birds living in the wild survive about three times as long as similar-size mammals do in all the care and comfort of a zoo or your house. The overall avian superiority in survival has some interesting wrinkles, though.

Seabirds, for instance, are particularly long-lived. Albatrosses, some of the largest of the seabirds, are suspected of being the longest-lived of all wild birds. Indeed, the longest-lived known-age bird alive today is a Laysan albatross (*Phoebastria immutabilis*) named Wisdom. Yes, when you have been around long enough and people know it, even wild birds acquire their own name. Wisdom lives—make that *breeds*—on Midway Island, a small atoll that sits, as its name implies, midway between Asia and the Americas. We know Wisdom's minimum age because in December 1956 an ornithologist aptly named Robbins—Chandler Robbins—gave her a ring. That is, he placed a numbered identification ring on her leg. Because she was incubating an egg at the time and because Laysan albatrosses don't typically produce eggs until they are at least five years old, Robbins figured she was born or, more properly, hatched no later than 1951, possibly even earlier.

Robbins was a vigorous thirty-eight years old at the time. Forty-six years later, still putting rings on birds at the age of eighty-four, Robbins caught her again. Stumbling upon a bird he had ringed half a century earlier amid the half million other albatrosses that breed on Midway Island seemed almost miraculous. Realizing that she now had to be in her fifties, much older than most researchers suspected wild albatrosses got to be, he alerted resident scientists that this was a very special lady. They have been keeping close track of her ever since.

To the uninitiated, albatrosses look something like seagulls with extra long wings. Wisdom's body, for instance, is about the size of a small cat, but she has the wingspan of LeBron James. With these long wings, albatrosses can glide above the waves for hour after hour without a single wingbeat. They are known to follow ships at sea for days at a time, and in the sailing ship era, killing an albatross was considered bad luck. You may recall that the ancient mariner in Coleridge's eponymous poem paid the price for killing an albatross by having its corpse hung around his neck, giving birth to a cliché for an unwanted burden that has lasted for nearly two centuries.

Wisdom, like other albatrosses, is poorly designed for take-offs and landings. Taking off requires a short run to pick up momentum, kind of like an airplane accelerating down a runway. Albatross landings, particularly if there is a bit of wind, sometimes remind you of the crash landings of ski jumpers who lose a ski while in the air.

Albatrosses not only have to survive these crash landings, but they have to survive several years living entirely at sea, flying thousands and thousands of miles, while they grow up. Once they reach that gawky adolescent stage, they return to their island home and typically settle down within sight of the spot where they were hatched. Now they have to survive a further couple of years of awkward courtship, perhaps *the* most awkward courtship this side of a prom night or a disco lounge, searching for a mate.

Once they identify Mr. or Ms. Right, they pair up for life and produce their single egg in most years. The pair take turns either guarding their chick or flying off to sea—sometimes for weeks at a time—to find those nutritious squid that they regurgitate to their chicks. By August, Midway Island falls strangely quiet. The half million squawking albatrosses are gone, foraging for the next few months at sea, rebuilding their energy stores for their next breeding attempt. Adults return to mate and jostle for nest sites around the beginning of December.

Wisdom has survived some feather-raising experiences, such as countless tropical storms and hurricanes. At the age of sixty, she even managed to survive the tsunami that rolled eastward from the 2011 earthquake that destroyed the Fukushima nuclear power plant in Japan and killed sixteen thousand people. That tsunami swept over Midway at midnight, killing more than a hundred thousand albatrosses but not Wisdom.

Through it all, Chandler kept tabs on Wisdom, right up to his death in 2017, when he was ninety-eight and Wisdom was at least sixty-seven years young.

Today, at age seventy or possibly a bit older, Wisdom seems as spry and energetic as ever, having fledged at least eleven chicks in the past twelve years. Wildlife officials on Midway calculate that over the course of her life she has flown more than 3 million miles, far enough to make six round trips to the moon. That didn't stop her from laying another egg recently (figure 4.2). She is still feeling good enough to have taken up with a much younger fellow

Figure 4.2
Wisdom, the oldest known wild bird, shown here tending her egg just before her sixty-eighth birthday. Note her identification band (right leg). Currently at least seventy years old, she is still churning out chicks.

several years back. Yes, she has outlived several life partners over the decades. Her newest beau has recently been given his own name—Akeakamai. In Hawaiian, that name appropriately enough means "lover of Wisdom."

What if we ignored the actual number of years lived and focused on longevity in terms of body size, using my longevity quotient (LQ), which, remember, is how long an animal lives relative to the record longevity of a zoo mammal of the same body size. When I previously said that *wild* birds as a group lived three times as long average *captive* mammals, it was another way of saying that wild birds overall have an average LQ of about three. Measured by LQ, Wisdom does not quite remain at the top of the avian longevity rankings—for now. I say "for now" because Wisdom is still going strong and we have no idea how long she might remain with us. Wisdom's LQ is 5.2. She has lived more than five times as long as an average zoo mammal the same size.

For now, then, Wisdom—representing the Laysan albatross—is only fifth in terms of LQ among bird species. The four species with higher LQs are all, like Wisdom, seabirds. That is, they spend their entire lives at sea, eat only seafood, and breed only on islands. In addition to island living, they all lay just a single egg per year and delay reproduction longer than most other birds their size. Delayed, slow reproduction is a feature of animals with exceptional longevity as we shall see again and again.

Another feature of long-lived species, in general, is that they occupy environmental niches or have physical designs that protect them from external hazards. Spending your life flying over the ocean except when diving into it to catch prey and breeding only on islands protects seabirds from a number of land-based dangers such as most predators and fire. In fact, if flight itself is a key feature that allows exceptional bird longevity, exceptional longevity even for a bird also seems linked to the safety of island life. Flight, despite its high energetic demands, allows birds to relocate long distances away if local conditions deteriorate. You could even think of the annual migration that many birds make as a form of adaptive relocation to better conditions. For birds that spend their lives on land, flight allows escape from land-based dangers such as small meat-eating mammals.

Topping the bird longevity quotient ranking for now, anyway, is the Manx shearwater (*Puffinus puffinus*), a 450-gram (one-pound) seabird. Shearwaters were named for their habit of skimming low over the sea, tilting their wings back and forth so they seem to shear the tips of waves. The Manx shearwater was named for the Manx language, which was spoken by natives of the Isle of Man in the Irish Sea, where there were massive breeding colonies. Manx shearwaters nest on small islands that they visit only at night. Like other long-lived seabirds, they delay breeding until the age of five to seven years and lay only one egg annually. How long can they live? So far, we know they can survive at least fifty-five years,[4] which because of their relatively small size gives them an LQ of 6. Perhaps most remarkably for a bird that lives so long, it migrates annually some six thousand miles from its breeding sites in the north Atlantic to winter in the waters off Brazil and Argentina. Well-known British ornithologist Chris Mead, who ringed more than 400,000 birds over his forty-year career, estimated that a Manx shearwater that reached age fifty would have flown over 5 million miles in its lifetime.

Land birds, despite the additional dangers they face, can also live an impressively long time. The highest known LQ for a land bird is for the mourning dove (*Zenaida macroura*). A sleek, light gray, fast-flying (up to ninety kilometers [fifty-five miles] per hour) gamebird, mourning doves thrive around human dwellings and are often seen sitting quietly on telephone lines or on the ground foraging for seeds that make up 99 percent of their diet. In contrast to long-lived sea birds, they can reproduce at one year of age. They, in fact, lay clutches of two to three eggs (a low number for a bird their size) but can raise two clutches of chicks per year. In the first few years of life, their mortality from environmental accidents, such as human hunters and animal predators, is high, but if they survive those dangers, they can survive a very long time indeed. Because of their abundance and interest by hunters, nearly 2 million mourning doves have been ringed in North America. More than 85,000 ringed birds have been reencountered, usually by the hunters who shot them. Despite the hunters, a record-holding mourning dove survivor is a male originally ringed in Georgia in 1968 and shot (of course) in Florida in 1998 some thirty years and four months later, giving these 130-gram (4.5-ounce) birds an LQ of 4.2.

Importantly for understanding natural patterns of longevity, not all bird species are long-lived. The general pattern is that those rare bird species that spend most of their time walking or running rather than flying—or are weak, occasional flyers—are relatively short-lived—more like *T. rex* or a captive mammal than an albatross.

Consider the turkey (*Meleagris gallopavo*). I don't mean the bizarre, domesticated Thanksgiving-eaten version that has been artificially bred for breasts so enormous and legs so plump that they are too heavy to fly and, in fact, can barely run. Consider instead the more regal, wild, and free version that Benjamin Franklin suggested might be a better national symbol than the bald eagle. Wild turkeys can fly, although that is not their normal form of locomotion. They typically fly in short bursts when being chased or when they flap their way into trees to roost for the night.

Like other birds and mammals, turkey muscles reflect their lifestyle. Turkey breast (that is, flight) muscle is white because it is adapted for short, intense bursts of energy and so needs little of the dark, oxygen-storing pigment, endurance-enhancing myoglobin that turns muscle dark. Turkey leg muscles are dark because they are specialized for endurance. Turkeys rely more on running than flight, and so the drumsticks are packed with myoglobin. Bird species that are long, strong fliers (like mourning doves and virtually all migratory species) have dark red breast muscle, or if you prefer, meat. Among birds, red breast meat indicates long life, and white the reverse.

Turkeys reach sexual maturity quickly for a bird so large. Males, or more formally "toms," run to about eight kilograms (seventeen pounds), and hens are about half that size. Hens lay roughly a dozen eggs over a two-week period, which will be mature adults by the next breeding season. So compared to seabirds, turkeys reproduce early and often. Not surprisingly, then, they don't live all that long. The oldest wild turkey ever recorded was a male found dead on the ground of unknown causes near New Salem, Massachusetts, in September 1992.[5] Its band identified it as being at least fifteen years old, giving this large, weak-flying species a longevity of 1.0—exactly the longevity you might expect from an *average* mammal in a zoo and by chance exactly the same LQ as a dog or a European rabbit. Turkeys breed like bunnies, and they seemingly die like bunnies, too.

WHAT WE MIGHT LEARN ABOUT HUMAN HEALTH
FROM BIRD LONGEVITY

Let me describe a mystery bird that vividly illustrates some of the biological challenges birds have had to conquer in order to live as long as they do.

My mystery bird is tiny, weighing about the same as a US penny. When active, it requires so much energy that it must eat up to several times its own body weight in food per day to avoid starving. During flight, its wings beat eighty times *per second*, and each gram of its flight muscles produces up to ten times the energy of the muscles of an elite human athlete when both are working at maximum capacity. In fact, it has the highest metabolic rate of any endothermic animal—and it is *very* endothermic. Its typical body temperature of 40°C (104°F) would be a dangerously high human fever. It requires so much fuel to maintain this level of energy expenditure that, when inactive, it drops its body temperature to that of the surrounding environment in order to not starve to death while sleeping. Its heart pounds away at a machinegun-like twenty-something beats per second. Even at rest, it breathes 250 times per minute, about the same rate as a panting dog, to get enough oxygen. Finally, its normal blood sugar concentration would make it dangerously diabetic if it were human. How long do you think it lives?

I have just described a hummingbird—specifically, a ruby-throated hummingbird (*Archilochus colubris*), the species commonly seen flitting around parks and gardens in the eastern United States. Hummingbirds fuel their frenetic lifestyle by sucking nectar—a rich blend of sugars—from flowers. They also suck up a few tiny insects along the way for protein. Their flight is a marvel to see and hear. Their name, of course, comes from the deep hum of their wings that beat so fast as to be almost invisible.

The 330 or so hummingbird species—all living in the Americas— are the only birds able to fly forward, backward, and hover. They can fly straight up and down like a helicopter and even do somersaults and other aerial acrobatics. Males, in particular, court females with displays of aerial acrobatics that seem impossible even as you watch them. On top of that, ruby-throated hummingbirds for all their tiny size and high energy demand fly six hundred miles *nonstop* across the Caribbean twice each year during migration to their tropical wintering grounds. If birds in general perform astonishing physical feats, hummingbirds' feats are among the most astonishing.

If that were all you knew about this hummingbird, you would no doubt expect them to be short-lived. After all, they live life in the fastest of fast lanes, and with few exceptions, animals that live fast, die young. But you would be wrong. Ruby-throated hummingbirds can live more than nine years in the wild, even as they perform their death-defying flights across the Caribbean twice per year. Nor are they the longest-lived hummingbird. The similar-size broad-tailed hummingbird (*Selasphorus platycercus*), with similar energy demands, can live up to at least twelve years in the wild.[6] Recall that the much bigger mouse, as previously noted, lives only a few months in the wild and only about three years as a well-cared-for pet. Therein lies a secret that once fully understood may help develop ways for people to remain healthy longer.

Hummingbirds are an extreme example, but virtually all of bird biology can be understood in terms of adaptations to the exceptional energy demands of powered flight. Those energy demands all suggest that birds should be short-lived, but they are the opposite. Their body temperature is higher than ours, their resting metabolism is up to twice as high as a mammal of the same size, and during flight, their metabolism cranks up even further. Even gliding flight such as performed by gulls, vultures, and albatrosses may look almost effortless to us but doubles or triples the birds' resting metabolic rate. Fuel for their exceptional energy demands is supplied by blood sugar levels that would signal uncontrolled diabetes in a human. Uncontrolled diabetes resembles accelerated aging more than virtually any other disease.

High energy, high heat, and high blood sugar should accelerate a number of the major processes that contribute to aging, one of which is free radical production. Recall that free radicals are molecules that can damage all classes of biological molecules, including DNA. To maintain cellular health, free radicals need to be destroyed rapidly by our antioxidant defenses, and the damage they inevitably cause needs to be repaired rapidly. Birds must have exceptionally effective antioxidant defenses and exceptionally rapid repair mechanisms. In fact, some of the few studies that have been done trying to understand bird longevity found that their cells produce fewer free radicals at the same rate of energy production as similar-size mammals. However, we don't understand how they do it. They also can withstand more free-radical damage before their cells die. We don't understand how they do that, either.

The other aging process that according to what we understand about aging should be accelerated in birds is browning of proteins. Proteins power the chemical reactions that define life. In their role powering chemical reactions, proteins need to be folded in complex and precise ways, like origami. Any slight deviation from perfect folding compromises their function. Imperfectly folded proteins not only lose function, but they become sticky, causing them to clump together with other misfolded proteins. The plaques and tangles of Alzheimer's disease are particularly well-known clumps of misfolded proteins, but there are many others.

Proteins misfold spontaneously in the chaotic, bumper-car environment of our cells all the time and are broken down and their parts recycled regularly. However, one particular type of protein misfolding bedevils slowly recycled proteins and is most relevant to birds and to diabetics. This is the browning reaction, which it is caused by heat and sugars. Sugars will attach themselves spontaneously to proteins, disrupting their precise folding. The higher the heat, the more concentrated the sugar, and the faster this browning reaction happens. It happens very rapidly at the temperatures we use in cooking. Meat and toast brown when heated because of this reaction. The same thing happens in our bodies, only much more slowly. For instance, our tendons and ligaments are composed of collagen, a protein that stiffens with age due to browning. Aging athletes have browning to thank for their increased risk of injury. Because of birds' higher body temperature and elevated blood sugar concentration, their tendons, ligaments, and other tissues should brown at a much higher rate than mammals. But they don't.

How birds prevent free-radical and browning damage is something from which human health could benefit. Do they have unique antioxidants that prevent free-radical damage? Do they have unique ways of degrading damaged proteins? They must also have mechanisms that preserve cellular functions in the face of life's challenges. There has been a little research on bird aging processes but never a large, sustained effort like we might have if they were being studied for cancer prevention. Medical research remains largely mired in the study of short-lived laboratory species such as fruit flies and mice, from which we may learn little to improve or extend human health. A Manhattan Project to understand birds' exceptionally slow aging and their ability to maintain strength and endurance throughout life would be a fine use of research dollars.

5 BATS: THE LONGEST-LIVED MAMMALS

I once had the good fortune to team-teach a field ecology course with Donald R. Griffin. Actually, I was the instructor of record, but Griffin, retired at the time and living nearby, asked if he could "tag along" on the course. To me, this was a little like Leonardo asking if he could "sit in" on your drawing course. Griffin was long famous for discovering (and naming) bat echolocation, the ability to "see" the world in considerable detail by listening to echoes of the ultrasonic shrieks they make. He had also pioneered studies of animal homing behavior and invented the field of cognitive ethology—animal behavior studies that assumed other animals, like people, were conscious, thinking beings.

Don told me a story during that course that has stuck with me. He and a group of college students were banding small bats—bats that weighed less than an American 25-cent coin—in a Vermont cave. Don had been working in that cave for years, and occasionally, he stumbled upon a bat that he had previously banded. When this happened, he called out the band number to a designated student note taker standing at the cave entrance, who recorded it and looked up in the field records when it had last been encountered. Calling out a number this particular time, Griffin waited for the student to respond. He waited . . . and waited . . . , and finally the student shouted, "Jeez, that one is older than I am."

BAT ORIGINS

Bats represent the fourth and last time that powered flight evolved in animals. We don't know exactly when and where they evolved or from what ancestors they arose. What we do know is that by 65 million years ago, numerous flying bat species existed, although these earliest species had not yet evolved the ability to echolocate. Recall the order and timing of powered flight origins. Insects took to the air roughly 300 million years ago, pterosaurs about 200 million years ago, birds about 150 million years ago, and bats only about 65 million years ago. Is it any wonder that competition with the already exquisitely developed flight of birds forced bats into the night to avoid losing the competitive battle?

We know so little about bat origins because their small delicate bones do not preserve well—especially in the humid tropics where we assume they originated. The first identifiable bat fossils already had well-developed adaptations to flight. Into the gap of uncertainty have leapt a series of speculations about bats' closest existing relatives. For many years, they were assumed to be most closely related to shrews, another group of small, nocturnal, insect-eating mammals. Another hypothesis suggested their closest relatives were the colugos, tropical gliding mammals sometimes known as *flying lemurs*, even though they are not lemurs and do not fly. One researcher even surmised on the most slender of anatomical evidence that bats were really flying primates—an idea that practically no one believed except himself. Modern molecular research indicates that bats are related to a large and diverse group of living mammals called the Laurasiatheres, which include everything from cows, deer, and horses to whales, hedgehogs, moles, shrews, cats, dogs, and bears. Which of these to choose as the closest bat relative is still up in the air, so to speak. However, the first two papers to provide complete bat genome sequences both inferred that the closest living relatives of bats are horses. Go figure.

MODERN BATS

Despite their sinister reputation, bats are, to my mind, the most remarkable of living mammals. They have been wildly successful in an evolutionary sense. Today, there are more than a thousand living bat species, accounting for about a fifth of all mammals. Despite these numbers, we know relatively little about only a few dozen species.

Bats live on all continents with the exception of Antarctica, and it is likely that 40 million years ago, when it still had a subtropical climate and formed one land mass with Australia, they lived there, too. Bats are often the only native mammals to inhabit oceanic islands. Not only are there lots of bat species, but there are lots of individual bats—enough that on summer evenings in the Texas Hill Country, massive swarms of bats emerging from caves are picked up on weather radar.

Although those of us living in the northern temperate world may stereotype bats as small, nocturnal, insect-eating cave dwellers, bats are much more diverse than that. In addition to the canonical caves, they roost in rock crevices, under loose bark, inside hollow trees, and in cave analogs like mines, barns, and attics. Some even sleep under leaves they modify into umbrellas. The smallest bat species is indeed a nocturnal, insect-eating, cave dweller not much bigger than a bumblebee, but the largest is a nonecholocating fruit eater about the size of a large seagull. Besides insects and fruit, some bat species eat flowers; others eat leaves; still others consume nectar, pollen, lizards, fish, frogs, and small mammals; and vampire bats subsist entirely on blood.

Bat echolocation, which Griffin discovered at a time when radar and sonar were the latest in highly classified military technology, is so exquisitely developed that bats can avoid objects as thin as piano wire in total darkness. They can identify, track down, and capture insects on the wing in the dark despite the insects' desperate evasive action. I have particularly enjoyed watching these aerial acrobatics in darkened Venezuelan cinemas, where moths attracted to the projector's light are chased by bats attracted to the moths with the drama playing out in a shadow play on the movie screen. In fact, I found this considerably more entertaining than the poorly dubbed B movies that were usually playing.

Bats invented massive mammalian crowds. Long before Rome became the first human aggregation of greater than a million individuals in the second century BC, bats packed together by the millions as tightly as any subway car at rush hour. Viruses and other infectious organisms love crowds. So millions of years of millions of bats crammed together in a confined space have made viruses and bats best friends, as we discover to our dismay again and again. When friendly bat viruses spill over into humans, they can become decidedly less friendly, as we know from rabies, Ebola, Hendra, Nipah, Marburg, SARS, and most recently, SARS-CoV-2, the coronavirus that causes COVID-19. Bats harbor more than eight hundred coronavirus species that have not become a problem—yet.

On the upside, millions upon millions of bats eat billions upon billions of crop-destroying and disease-transmitting insects. Millions upon millions of bats also produce tons upon tons of guano (how many other mammals have a special word for their excrement?), formerly used to make gunpowder and other explosives but best known as the crop fertilizer crucial to the nineteenth-century development of high-intensity farming. Bat guano also provides the nutrients to support whole ecosystems of cave organisms—from fungi to fish, so to speak.

How Long Can Bats Live?

Compared to birds, we know relatively little about bat longevity, but what we do know is astonishing. If understanding longevity in wild animals is a numbers game—that is, you need to mark and, more important, recapture multiple times, lots and lots of individuals to be confident that you have learned something about a species' capacity for longevity—then it is easy to understand why we know so much less about bat than bird longevity. Legions of professional researchers and amateur birdwatchers have banded and recaptured millions upon millions of birds for a century. By contrast, bat researchers are rare, although singularly dedicated. They need to be willing to become nocturnal themselves for months at a time and traipse and wade through forest, field, stream, and cave in the dark—all to investigate creatures of which most people are dubious at best and terrified at worst. Even their enormous colonies can make bat longevity difficult to study. Imagine banding a thousand bats—no mean feat—that emerge from a cave occupied by several million bats. What are the odds that you would recapture those same bats, much less capture them often enough to get some idea of how long they live?

So most of what we know about bat longevity is accidentally acquired knowledge. Typically, researchers looking for bats stumble upon a colony of bats that they or, more often, other researchers have banded years before and discover that some of the banded bats are still alive. Given all these difficulties, our estimates of how long bats live in the wild are likely considerable underestimates, especially for species that have not been monitored extensively.

The little brown bat (*Myotis lucifugus*) provides a classic example. This is the most common bat species in much of North America and the species in which Don Griffin discovered echolocation. During warm months, they can be found during the day roosting—that is, sleeping while hanging by their feet (a bat specialty)—in and around buildings, inside hollow trees, and under rocks or piles of wood. At night, they fly miles foraging for insects. During the winter, they hibernate in caves or abandoned mines.

Over the winters of 1961 and 1962, two professors, Wayne Davis and Harold Hitchcock, and their students from Middlebury College in Vermont banded nearly ten thousand hibernating little brown bats in an abandoned iron mine in eastern New York. No one returned to that mine until the early 1990s, when bat conservationists, checking the mine for bat presence, discovered, to their surprise, a few of the originally banded bats were still alive. In thirty years, the bats had remained faithful to their hibernation site. Returning in subsequent years to continue their survey, the conservationists discovered that little brown bats, animals one-third the weight of a mouse, could live at least thirty-four years amid all of nature's challenges.[1] Emphasis here is on "at least." The ages of the originally banded bats were unknown. Placing this exceptional longevity in a larger context, little brown bats—all ten grams (0.4 ounces) of them—have a longevity quotient of *at least* 7.5, considerably greater than any wild bird.

It is important to note a general trend here. Possibly due to the rigors of flight, the major milestones of bat life history—sexual maturity, rate of reproduction, and death—are reached slowly compared to other tiny mammals. Instead of having five to seven pups every couple of months and reaching reproductive age at as early as two months of age like mice, most bats have one pup at a time (flying while pregnant having its special challenges), give birth once per year, and have pups that require close to a year to become reproductive adults. Maybe that makes it less surprising that bats are especially long-lived, although it doesn't tell us how they do it.

We might learn a bit more by considering a few individual species, starting with possibly my favorite of all bat species, the common vampire bat.

Common Vampire Bat (*Desmodus rotundus*)

Vampire bats got their name from the Count and not the other way around. Actually, Bram Stoker's 1897 novel *Dracula* did not give them their name directly. The vampiric folk tales of malevolent blood-drinking undead that Stoker adopted for his novel did. When actual nocturnal, blood-drinking animals were brought to Europe from the New World (Eurasia has no vampire bats), the naming opportunity was too good to pass up.

Vampire bats are the size of a mouse. They live in the New World tropics and subtropics, from northern Mexico to northern Argentina. Like their fictional namesake, they spend the daylight hours roosting in the darkest places they can find—caves, old wells, hollow trees, boarded-up abandoned buildings. By night, they awake and forage for blood. Yes, their adult diet is 100 percent blood, although vampire pups drink their mother's milk like the rest of us mammals. There are three species of vampire bats. Two of them drink the blood of birds—wild birds for much of their evolutionary history but mainly chickens these days. The common vampire drinks the blood of only mammals. Opportunistically, they may dine on human blood, but their most common hosts are cattle and horses, probably because cattle and horses often sleep out-of-doors at night and do not have hands to slap away bloodthirsty bats. In fact, rabies transmitted from the saliva of vampire bats is a significant agricultural problem for cattle ranchers in Central and South America.

Rabies can be transmitted by many bat (and other mammal) species, but vampire bats have the reputation. In reality, only a small percentage of vampires carry rabies because despite what you may have heard, rabies kills them and other bats just as it kills humans, skunks, and raccoons. Knowing this, when I once got bitten on the finger by a vampire bat while removing it from a mist net, I realized it could be a problem. As I was far from any medical help at the time, I had to decide whether to interrupt my research for several days, drive several hundred miles over spine-jarring roads in search of rabies vaccine, or play the odds and assume that the bat that bit me either wasn't infected or hadn't transmitted enough virus to kill me. Balancing interrupting my research versus a small chance of dying an agonizing death from rabies and the possibly more painful prospect of admitting to my wife that I was dying an agonizing death because I didn't want to interrupt my research for a few days, I went for the rabies vaccine.

Making a living by drinking the blood of much larger animals has some special challenges. First, vampires have to find sleeping mammals on which they can alight and creep about without alerting them. Vampires, unlike most bats, have fore and hind limbs that allow them to creep, walk, run, or hop as necessary. At a distance, you could easily mistake a vampire bat scuttling around on a cow for a large spider. If successful at not disturbing their prey, they use infrared sensors in their nose to find a spot where blood flows near the skin. With specialized teeth, they shave away the fur to expose bare skin, cut a hole about the diameter of a pencil through the skin with—as I can attest—razor-sharp incisors, and begin lapping the blood as it flows. Their saliva contains chemicals that dilate the local blood vessels to quicken the flow as well as anticoagulants to keep it going.

There is a reason that relatively few animals specialize in a diet of blood. It is about 90 percent water, the rest being virtually all protein. Blood may be the ultimate low-calorie, high-protein diet. Low-calorie diets may be fine for sedentary humans, but for wild animals with high energy demands, a low-calorie diet means starvation is always only a few days away.

Because of the low calories, vampires need to drink a lot of blood. A typical feeding bout lasts about thirty minutes, during which they consume about 60 percent of their body weight in blood. As you might have guessed, this extra water weight poses a potential problem for a flying animal. They solve this problem by having the ability to jettison this extra water weight rapidly. Within a couple of minutes of starting to drink blood, they start to urinate, dumping the extra weight almost as fast as they take it on. Still, by the time they are finished, they weigh about 20 percent to 30 percent more than when they left the roost. They now labor into the air, return to the roost, and spend the rest of the night digesting their hard-won meal.

I have lingered on the vampire's blood diet because it is a key to understanding a lot of their biology, including their longevity. We know a great deal more about their biology than we do about most bats because it turns out they do well in captivity and are economically important to tropical cattle ranchers.

Their almost exclusively protein diet means they have virtually no ability to store energy as fat. Consequently, vampire bats will starve to death if they fail to eat for as little as seventy-two hours. As they typically avoid foraging on bright moonlit nights, probably to avoid owl predators, vampires are always teetering on the edge of starvation. Evolution has provided them a nice insurance policy, though—the other bats with whom they roost. Although they might roost in colonies of up to hundreds of other vampires, within those colonies there are smaller clusters of a dozen or two animals with especially chose social ties. Within those clusters, they share food as needed. I'm not sure whether it will warm the cockles of your heart as it should, but well-fed vampires will regurgitate blood to less fortunate cluster-mates who have been unsuccessful at foraging.[2] As some bats are successful on some nights and others on other nights, there is reciprocal blood sharing over time. These groups are largely (but not exclusively) formed of several mothers and generations of daughters, so sharing is often (but not always) with, yes, blood relatives. What really determines which individuals share food is whether they are bat buddies. *Bat buddies* are what I call those who spend a lot of time together and mutually groom one another, especially those who have previously shared blood with them. Bats with numerous blood buddies have multiple food insurance policies.

So how does blood dining affect the life course of vampires? It means that energy is severely limited. Of necessity, they do everything a bit slower than most bats, and remember all bats march through life's milestones slower than most other small mammals. For instance, a mouse gestates its five to seven pups for three weeks, and a standard issue bat does so for a single pup for three to six months. A vampire bat requires even more time—seven months—before its single, large pup is born. Bats as a group, I should note, have the largest babies relative to maternal body size of any other mammals. A newborn mouse weighs about 5 percent as much as its mother, and a newborn vampire about 25 percent of its mother's weight. Mouse pups nurse for about three weeks and are weaned at half of their adult size. Most bats nurse their single pup for three to six months until they are almost adult size, but vampire nurse theirs for eight months. To give them a taste of adulthood, so to speak, vampire mothers begin regurgitating blood to their pups before they are fully weaned.

Oh yes, and now that we get to it, how long do these mouse-size vampires live? In captivity, females live up to thirty years, and in the wild, the longevity record so far is eighteen years. Males are a bit shorter-lived. Remember that longevity quotient is calculated based on the longevity record of captive mammals, so directly comparing it to other mammal species, vampire bat LQ is 5.5. That is, they live more than five times as long as an average mammal. Even in the wild, they live more than three times (LQ = 3.25) as long as a same-size, average mammal in a zoo. Note that this is close to the wild longevity of a house sparrow. Do birds and bats share similar longevity secrets? We will come back to that.

Let's now take a look at a very different kind of bat—one that lives long in absolute terms, even when not adjusting for small bat body size.

Indian Flying Fox (*Pteropus medius*)

Flying foxes are about as different from vampires or little brown bats as can be imagined. Instead of the pushed-in face, beady eyes, enormous ears, and menacing teeth, flying foxes are cute as any puppy. Flying foxes are also considerably larger than most other bats, ranging from ten to fifty times the weight of a vampire bat. Instead of spending the day hidden away in the dark refuges, flying foxes hang conspicuously from the branches of tall trees, looking like so much dangling fruit at first glance. Those in large colonies, which may number in the thousands, constantly jostle, chatter, and from time to time take off for lazy forays around the roost searching for a new resting place with better neighbors. At dusk, they lift off *en masse* and disperse to their individual feeding grounds, which may be an hour or more distant as the bat flies. They return to their roost just before sunrise to sleep, jostle, chatter, and complain for another day. When I worked in Papua New Guinea, the flying foxes roosting in the palm trees around the airport of the seaside town of Madang were so numerous and active at certain times of day that flight schedules had to be altered to minimize the risk of plane-bat collisions—events that would be unfortunate for both.

Instead of hunting down elusive prey by sound, flying foxes use their large eyes, excellent night vision, and sensitive noses to locate fruit, their prey of choice. In the tropics, they provide major services to forests, dispersing fruit seeds and pollinating flowers of many plant and tree species. Their roost trees also benefit from the guano fertilizer they provide in abundance. On the downside, their fruit-eating habit does not endear them to fruit farmers, who think of them legitimately enough as crop thieves. Also, like other bats who for millions of years have lived in large colonies, they carry plenty of viruses, some of which (like the Hendra and Nipah viruses) lethally spill over into domestic animals or people from time to time. Often, the spillover is *from* domestic animals *to* people. An outbreak of Nipah virus transmitted from Indian flying foxes to pigs to people in Malaysia in 1998 killed more than a hundred people, mostly men working on pig farms, which led to the preventative slaughter of more than a million pigs. To date, Nipah virus has not been able to spread efficiently from person to person, so despite at least eight outbreaks in the past twenty years, its human impact remains small. But the possibility always exists that just the right mutation combined with just the right situation could set off another global pandemic, as we discovered to our regret in 2020.

The Indian flying fox is one of the largest species of flying foxes, which makes it one of the largest bats (figure 5.1). About the size of a seagull, they range across the Indian subcontinent from Pakistan and Bhutan to Bangladesh and down the Malaysian peninsula, living in large colonies on tall thin trees near bodies of water and agricultural fields. Although they are called fruit bats and do eat many types of ripe fruit (including figs, mangos, guavas, bananas, almonds, dates, and a host of forest species that people do not eat), they also eat flowers and drink nectar. More than three hundred species of plants used for around five hundred economically valuable products rely on them to disperse their seeds.

Because of their agricultural significance as well as their role in transmitting Nipah virus, Indian fruit bats have been studied more than most bats both in the wild and in captivity. From those studies, we have learned a lot about their development and reproduction, including arguably more than we need to know about their sex life.

Figure 5.1
Indian flying fox with pup. The rigors of flight while carrying large pups during pregnancy or even after birth combined with the extra food demands to feed a growing pup may play a role in male bats often being the longer-lived sex.

Bonobos, who along with chimpanzees are our closest primate relative, have a reputation for rampant sexual free-for-alls. Indian fruit bats are the bonobos of bats. Biologists call their mating system *polygynandrous*, a dull, Greek-rooted scientific term meaning females may mate with any number of males and males with any number of females. More attention-getting, perhaps, males and females regularly engage in oral sex, often but not always as a prelude and an encore to actual mating. A paper title that I confess I never expected to come across in a scientific journal revealed that "Cunnilingus Apparently Increases Duration of Copulation in the Indian Flying Fox." Yes, after intently monitoring nearly sixty sexual bouts, researchers learned that ten extra seconds of cunnilingus foreplay apparently rewards a male with a lengthy seventeen seconds of copulation rather than the brief fifteen seconds a less enthusiastic male is allowed.[3]

More prosaically, like most other bats, females typically have a single young born after a five-month pregnancy once per year. The pups weigh only about an eighth as much as their mother at birth, which is about half the relative weight of most bat newborns. Indian flying fox mothers compensate for this early birth with extra maternal care, carrying their pups 24/7 for the first few weeks of life, until they are big enough to be parked in the roost tree while mom goes to work foraging. At about three months of age, when it has grown rapidly to about 90 percent of its adult size, the pup begins to fly on its own. When it is five months old, the mother cuts off its access to milk, and it is on its own. It will be ready to perform its own sexual antics and perhaps produce its own pup by the time it is two years old. In zoos, where flying foxes eat more and work less than in the wild, they grow faster and may reach sexual maturity in as little as a year.

In terms of longevity, little is known about their longevity in the wild, but Indian flying foxes can live at least forty-four years under the right captive conditions, making them in calendar years, the longest-lived bat species we know.[4] That forty-four-year-old male flying fox had quite the odyssey. Born in the wild in India in 1964, he was captured and shipped to the Milwaukee County Zoo a year later when still prepubescent. At the age of twenty-one, he moved to a warmer climate—the famous San Diego Zoo— where he spent the rest of his life in the California sun.

I should point out that it isn't unusual for zoo animals to move from one zoo to another, perhaps even move several times if they live a long time. Zoo conditions are constantly changing due to budgetary constraints, social constraints (social species may require companionship), or overpopulation of fecund species. Zoos also need to match their menagerie with their display facilities, and if they are breeding a species, they may need to avoid inbreeding. It is during such transfers between zoos that birth records most often get switched or confused. The result can be exaggerated longevity. The false claims of a 157-year-old elephant and a 147-year-old cockatoo no doubt came from mixing up birth records.

In the case of the Indian flying fox, it is reassuring to find that other members of this species are reported to live at least into their thirties in other zoos. A female born in the London Zoo lived her entire life there, dying at the ripe old age of thirty-one years. Another male, named Michael, died peacefully at the age of 33½ in the Houston Zoo, after being born in Tulsa, moving to Oklahoma City as a ten-year-old, and making a final move at age twenty-three to spend his golden years in Houston.

Putting flying fox longevity in context, the longevity quotient of the Indian flying fox is 4.1, meaning it lives almost four times as long as expected for an average captive mammal of the same body size. No other similar-size captive mammals of any species approach forty-four years of life. For instance, a same-size black-tailed prairie dog (*Cynomys ludovicianus*) reaches eleven years, and the ringtail (*Bassariscus astutus*), a smaller relative of the raccoon, makes it only to sixteen years. The nonbat species of similar size with an LQ closest to the Indian flying fox is everyone's favorite the three-striped owl monkey (*Aotus trivirgatus*), a nocturnal primate of tropical America. The Methuselah of three-striped night monkeys was a male who was born, lived, and died at the age of thirty in the Prague Zoo—an LQ of 2.8. As we shall see later, primates as a group tend to be long-lived for their size, just not as long-lived as bats and especially not as long as Brandt's bat.

Brandt's Bat (*Myotis brandtii*)

Hagrid discovered the longest-lived wild bat. Hagrid, of course, is the giant, forest-dwelling gamekeeper of the Harry Potter novels. Hagrid wasn't the bat-catcher's real name. His real name was Alexander Khritankov, but I thought of him as Hagrid. He was and possibly still is a biologist at the Stolby Nature Reserve in central Siberia. Stolby is a 47,000-hectare (180-square-mile) nature reserve set aside by Joseph Stalin in 1925. Stolby is famous for its spectacular rock pillars and limestone caves. It was in one of these caves that Alexander found the Methuselah of wild bats.

I had come across a report in a bat specialists' journal of particularly long-lived bats found somewhere in Russia. As my colleague Andrej Podlutsky was a native Russian speaker, I asked him to see if he could track down the author of that report to find out more. He managed to connect with Alexander by email, and they arranged to talk if Andrej would ring the main reserve number at a certain time on a certain day. When not expecting a phone call, Alexander was roaming the taiga. One time, we learned he was out fur hunting to fund the purchase of a new computer. I always imagined him emerging from the forest in the dead of night, animal carcasses slung over a shoulder, to pick up a ringing phone outside a lonely cabin under the only streetlight for miles along a desolate Siberian backroad. Okay, it probably was nothing like my fantasy, but Andrej and Alexander did connect, and this is what we learned.

For several years in the early 1960s, one of the Stolby reserve biologists banded about fifteen hundred Brandt's bats. It was the height of the Cold War, around the time of the Cuban missile crisis. Russian science was not particularly focused on bats or bat conservation. Some twenty years went by before anyone entered those caves again. Sixty-seven marked bats—all males—were still there. Few visited the caves in the 1990s until Khritkanov rediscovered the marked bats again in the early 2000s.

Myotis, the bat genus to which both Brandt's bat and the little brown bat belong, sometimes doubles as their common name. That genus contains more than a hundred species living in almost every habitat in almost all climates on almost all continents—except (as always) Antarctica. *Myotis* are almost all small, echolocating bats that spend the night hawking insects out of the air with occasional pauses to rest and digest their prey. A few aberrant *Myotis* species are different. They specialize in scooping small fish from the water surface. The smallest species weighs about 2.5 grams (one-tenth ounce), making it a competitor with some of the hummingbirds for the smallest endothermic animal alive. The largest species weighs about fifteen times as much—mouse size. The longevity records for wild *Myotis* species sprawl all over the place—black myotis seven years, fish-eating myotis twelve years, and, as we previously learned, little brown myotis thirty-four years. How much of this variability is real and how much represents lack of information about the presumptively shorter-lived species isn't yet clear. What is clear is that Brandt's myotis tops them all.

The last time Khritnikov spotted the last survivor from that 1962 banding expedition, it was *at least* forty-one years old.[5] And then . . . poof . . . : it was gone. No corpse, no good-bye, no nothing. Forty-one years for the tiny six-gram (one-fifth-ounce) Brandt's bat gives it the remarkable longevity quotient of precisely 10.0. Think about that. A bat small enough to be mistaken for a large butterfly in flight manages to avoid predators, survive famine, flood, pestilence, heat waves, and cold snaps decade after decade. To survive nature's continuous challenges, Brandt's bats as they age need to maintain the stamina to fly many miles each night and the agility to snatch insects, many of them taking desperate evasive maneuvers, out of the air every few seconds to avoid starvation. Human athletes in sports requiring strength, agility, or endurance never maintain their highest performance level for forty years.

As echolocating hunters, they must also preserve their high-frequency hearing. Hearing loss for an echolocating bat is a death sentence. High-frequency hearing is the first sensory faculty humans lose. Some English shopkeepers have taken advantage of this feature of human aging to discourage teenagers from loitering outside their shops. They do this by blaring annoying high-pitch buzzing tones that only children and teenagers can hear. Adult customers are oblivious. As they age, female bats in particular must retain an excellent spatial memory because after mothers forage many miles each night in the dark, they must be able to return to the precise roost where they parked their pup. No where-did-I-leave-the-car-keys moments are allowed.

One question bedeviled Andrej and myself. Why might Brandt's bat survive 20 percent longer than the little brown bat, which from every practical angle *is* the Brandt's bat of North America? One plausible answer may be climate.

HIBERNATION AND AGING

Small endothermic animals—mammals and birds—face a chronic energy challenge that only worsens with cold weather. Remember small size itself—because of the relatively large body surface (over which heat is lost) compared to its small heat-generating mass that small size necessitates—makes for rapid heat loss. To preserve a constant high body temperature, small animals must produce enough heat to compensate for this rapid heat loss. This is why small birds and mammals have higher metabolic rates than larger animals. The challenge worsens in the cold because the greater the difference between body temperature and environmental temperature, the faster the heat loss. So as nights get colder and longer, even more heat-producing fuel is needed. For an insect-eating bat as winter approaches, fuel—that is food—becomes increasingly scarce. Flying insects virtually disappear, particularly at the coldest time of day, when bats would be foraging. At some point, it becomes energetically untenable for a small bat to carry on. They give up on activity and maintaining their typical mammalian body temperature. They retreat into caves or other safe refuges and hibernate.

Hibernation, a mammalian specialty, involves allowing body temperature to drop by a variable but controlled amount to conserve energy. It is typically a response of small mammals like chipmunks and prairie dogs to wintery climates, although a few larger mammals hibernate as well. Bears, for instance, because of their large size and well-insulated dens, drop their temperature only a few degrees when hibernating, which is enough to reduce their resting metabolism to one-fifth its normal rate. Bears can still move during hibernation, although sluggishly. They seem largely oblivious to their surroundings when doing so, though. I discovered this when my wife talked me into entering a cage with a hibernating grizzly bear being studied at her veterinary college. To my dismay, soon after we entered, the bear lumbered to its feet and walked a few steps before flopping back asleep on the other side of the cage. Thankfully, it paid no attention to the terrified man a few feet away trying his best to look invisible.

Hibernating bats also hibernate variably, depending on local conditions. Brandt's and little brown bats live where winters can be brutally cold. While hibernating, they drop their body temperature to near freezing, which reduces their metabolism to less than 1 percent of its normal resting rate. One of the reasons, in addition to safety, that they hibernate deep in caves is that deep-cave temperatures are close to the *annual* average temperature of their environment—cold but not quite freezing—even in the depths of winter. Mammalian hibernators will die if they freeze, as ice crystals form inside their cells, bursting the cells apart. Exploding brain, heart, and other cells is not conducive to long life.

So getting back to the comparative longevity of little brown bats and Brandt's bat and its relation to the climate, in Essex County, New York, where the thirty-four-year-old little brown bat was discovered, winter is shorter and milder than in central Siberia, although in neither place is it particularly mild or short. For instance, during the coldest month—January—the average daily low temperature in Essex County is a brisk −14°C (8°F) in New York, but in Stolby, it is an even more brisk −20°C (−4°F). Due to this climatic difference, little brown bats hibernate in eastern New York for about six months, whereas Brandt's bats in Stolby hibernate for nine months. If metabolism—the fire of life—plays a major role in aging, maybe hibernation represents something like a "time out" from aging. Brandt's bat conceivably may be aging for only three months every year, whereas little brown bats age for six months. Mice in the laboratory get no timeout at all.

Cleary, this is a partial explanation, at best. Being capable of flight regardless of whether or not you hibernate matters whether you are a bird or bat. Birds do not hibernate (with the exception of one species, the common poorwill of the southwestern United States), yet most birds live much longer even in the wild than most mammals do in captivity. Among long-lived bats, neither the Indian flying fox nor the common vampire bat hibernate, but they still live four to five times longer than other mammals of their size. Hibernation may give an extra longevity boost, but something additional is going on with birds and bats. In a recent analysis of the comparative longevity of nearly a hundred bat species, University of Maryland biologists Gerald Wilkinson and Danielle Adams found that hibernating bats, on the whole, lived longer than nonhibernating bats, and the greater length of hibernation (estimated from the latitude at which they lived), the longer they lived.[6] Vampire bats, incidentally, may not hibernate, but they do enter shallow torpor between feeding bouts.

Hibernation makes me think of another astonishing feat of bat biology—their resistance to withering muscles with inactivity. When our muscles are not used, they wither quickly, as anyone who has worn a cast knows well. The rate of muscle loss associated with inactivity accelerates as we get older. Older people can lose as much as 16 percent of their lower body strength after only ten days of bed rest. It takes younger people a month or more to lose the same percent of strength.[7] Consider now bats that hibernate for months—as long as nine consecutive months. How much strength and muscle do they lose during that time? Virtually none.[8] At the end of that time, they wake up and fly away. How do they do it?

There are many aspects of bat biology related to their long lives about which we would like to know more. This is one of them. Another is how they manage to coexist with all those viruses. Some researchers have speculated that bats' immune systems, powerful enough to resist all those viruses, play a major role in their exceptional longevity.[9] We would also like to know how they deal with) free radical and browning damage to proteins year after year after year. After all, they have the same challenge of exceptionally high energy demand during flight that birds have. Like birds, they produce fewer free radicals for the same amount of mitochondrial energy production and manage misfolding of their proteins better than other mammals, but we don't understand how they do those things.

Another thing that I mentioned only in passing is why and how male Brandt's bats live so much longer than female bats. We will return to the issue of sex differences in longevity. For now, recall that all sixty-four Brandt's bats that lived longer than twenty years were males. The longest-lived little brown bats were all males, too. It may be that raising the largest pups compared with the size of the mother has its costs, or it may be something else. Among other mammals, females are often but by no means always the longer-lived sex. Can we learn something about aging in general by studying these sex differences?

So we have a great deal to potentially learn about aging from studying bats in addition to the secret of their long life. How do they maintain their hearing? How to they keep their muscles strong during months of inactivity? How do they maintain their endurance and agility for decade after decade after decade? How does their immune system cope with all those viruses? How do they remember exactly where, among millions of bats crammed into a cave, they parked their unweaned pup? And how can they fly dozens of miles in the dark and find their way back to the same cave in the first place? There is just beginning to be a major effort to understand bat longevity. A number of bat species have now had their genomes sequenced,[10] which may give us hints about where in their cellular biology to look for answers to the questions above. However, genome sequencing is only a start. It is something that we are good at and can now be done quickly and relatively cheaply. But looking at the genome can only point us in the direction that more focused cell biology and physiology will be required to truly understand bat longevity. As with birds, a Manhattan Project involving large teams of researchers focused on bat biology would be money well spent if we wish to discover nature's secrets for substantially extending health and well-being.

BATS, BIRDS, AND HUMAN HEALTH

Bats and birds differ from all the other long-lived animals in this book. They differ in that they live fast but live long. One of the routes often used for longer life is living slow. That is, if the basic processes of life proceed rapidly, the damaging side effects of these processes generally overwhelm animals early, whereas slowing down these processes slows down the damage allowing for longer life. Bats slow things down when they hibernate, although they live fast the rest of the time. Ectothermic animals live slowly almost all the time. However, living long while living fast is presumably the type of long life most humans would like. How many of us would want to live a longer life if the extra years have to be spent napping? We also want to extend our health, not just our existence. The fact that long-lived birds and bats preserve their physical strength, endurance, and agility and also maintain the acuity of their senses and their cognitive abilities to near the end of their exceptionally long lives is something that humans would like to emulate. Yet until we make a firm commitment to study in depth the animals that can do these things and not remain stuck in our research on the short-lived, rapidly aging species that fill our biomedical labs today, we are unlikely to make much progress in achieving longer, healthier human lives.

II LONGEVITY ON THE EARTH

6 TORTOISES AND TUATARAS: LONGEVITY ON ISLANDS

Charles Darwin first set eyes on giant Galápagos tortoises in September 1835. So far as we can tell, he gave no thought as to how long these behemoths might live in order to achieve their great size. Their size, he reported, was such that six to eight men were sometimes needed to lift one. His attention instead was drawn to their great numbers on the islands, their use as a plentiful supply of meat, and the distance and speed with which they would hike from the arid island lowlands to drink from freshwater springs on the more lush upper slopes. A trip of eight miles, he calculated, could be accomplished by a speeding tortoise in as little as two days if it traveled day and night.

He did note in passing that local inhabitants reported they never found a dead tortoise without it being evident that it had accidentally fallen from a cliff. From a modern-day, aging perspective, we might suspect that failing eyesight was responsible for these fatal accidents and wonder why they were never found dead from more obscure causes.

Turtles and tortoises developed a reputation for longevity millennia ago. By the way, if you are confused about the difference between a turtle and a tortoise, there is a good reason. The terminology differs by the species of English you speak. In Britain, the word *turtle* is seldom used at all. A tortoise is usually any land-dwelling reptile with a protective shell and a beak instead of teeth, whereas those that spend part or all of their lives in water are terrapins. In Australia, a tortoise is a freshwater turtle, possibly because Australia has no land-dwelling reptiles with protective shells and beaks instead of teeth. In America, the official word of the American Society of Ichthyologists and Herpetologists—and who could argue with such an official-sounding group—is that *turtle* is the generic term for any reptile with a shell and beak, regardless of whether they spend their time on land or water, and *tortoise* is reserved for slow-moving terrestrial turtles. Topping off the confusion, tortoiseshell jewelry was fashioned from the shells of hawksbill sea turtles before it became entirely synthetic.

Whatever you call them, turtles and tortoises probably achieved their longevity reputation because some were long ago adopted as pets and people noticed that it took them a very long time to become adults. The length of time it takes to reach adulthood is a crude indicator of how long something may live.

Moreover, once turtles did become adults, people noticed that they tended to live longer than their owners could remember. I should note right here that although I focus largely on the longevity of giant tortoises, all or at least most turtles and tortoises seem to take a long time to become adults, and once adult, they tend to go on and on. For instance, the widespread saucer-size painted turtle (*Chrysemys picta*) of North America takes a decade to reach adulthood, much longer than any bird or bat, and can live as long as sixty-one years in the wild. The slightly larger—let's call it dinner-plate-size—Blanding's turtle (*Emydoidea blandingii*) takes fifteen to twenty years to become reproductive, and one is known to have lived at least seventy-seven years in the wild.[1] Both of these species hibernate in the parts of their range that experience a cold winter, and both longevity records come from populations that hibernate. So to the extent that hibernation may be a "time-out" for aging, this might explain some—although not all—of their exceptional longevity.

However, the giant tortoises native to the Galápagos Islands and Aldabra atoll do not hibernate, and they clearly live an exceptionally long time. If small turtles like those mentioned above can live into their sixties or seventies and the general pattern is that larger species live longer than small ones, it seems reasonable to assume that Galápagos tortoises' great size suggests exceptional longevity indeed. Size may be part of the reason, but an equally plausible reason to my mind is that they evolved on islands. Let me explain.

THE STRANGE BIOLOGY OF ISLANDS

Ever since Darwin, islands have provided instructive lessons in evolution. That is because the biology of islands is bizarre, particularly the biology of oceanic islands. Oceanic islands are just what they sound like—islands that are far from any mainland, arose from volcanic eruptions on the ocean's floor, and over millennia piled up enough lava to eventually rise above the sea surface. As they typically form when one of the earth's tectonic plates creeps over a hotspot—a place where molten magma regularly bursts through the earth's crust—oceanic islands usually occur in chains, representing a time series recording when the plate crept over the hotspot.

A familiar example is the Hawaiian island chain. Midway island, or more properly Midway atoll, is one of the northwestern-most Hawaiian islands (and home to Wisdom, the world's longest-lived wild bird). Midway inched over the Hawaiian hotspot about 28 million years ago as the Pacific tectonic plate journeyed northwest (and as it continues to do today at about the rate that fingernails grow). As islands move away from the hotspot that formed them, they begin to erode and also gradually settle back into the sea from their own weight, sinking lower and lower until nothing of the original volcano may remain visible except perhaps bits of its fringing reef that surrounds its one-time rim. Now you have an atoll. On top of solving the puzzle of how evolution worked, Charles Darwin also solved the mystery of how atolls form while on his voyage aboard HMS *Beagle*. So reconstructing the history of the traditionally defined Hawaiian islands—with the oldest islands in the northwest and the youngest in the southeast—we know that Kauai emerged from the sea about 5 million years ago, Oahu 3 million years ago, Maui one million years, and the youngest, the big island of Hawaii, is just now finishing up passing over the hotspot. A soon-to-be new island— the Loihi seamount—is building toward the ocean surface about twenty miles southeast of the Big Island. We should see Loihi emerge from the sea in somewhere between ten thousand and a hundred thousand years from now.

The key biological feature of oceanic islands is that they emerge from the sea utterly devoid of terrestrial life. Their web of plant and animal life assembles haphazardly over time as chance determines the order in which new species arrive. Because of this haphazard arrival, species might encounter empty ecological niches to fill that on the mainland would have been filled millions of years ago. So island animals frequently evolve traits very different from their mainland ancestors. One way they often differ is in body size. There are island giants and island dwarfs of many species, including humans.

The flightless elephant birds of the island of Madagascar are an example of island gigantism (figure 6.1). Elephant birds were up to five times the weight of a large ostrich. Other island giants were the extinct moas of New Zealand, which would tower over the heads of the largest elephant bird but were more slender. Dodos, living on Mauritius, an island just east of Madagascar, were thirteen-kilogram (thirty-pound) flightless pigeons, about forty times bigger than city pigeons. The advantages of flight are considerably reduced on islands, and to the extent that being blown off the island in a gale is a problem, flight can even be a hindrance. So island birds and many island insects have lost the ability to fly. New Zealand's giant wētā is a flightless cricket the size of a mouse.

Figure 6.1
The elephant bird (*Aepyornis*), an extinct island giant of Madagascar, compared to ostrich, human and chicken. The also extinct giant moas (not shown) of New Zealand were considerably taller but not as heavy as the elephant bird. All giant island bird species became extinct in historical times.
Source: Drawing by De Agostini via Getty Images.

Another key insular feature is that top predators are often absent, the prey base being too small to sustain viable populations of predators, robbing flight of another advantage—the advantage to fly away from danger. Lack of large predators also accounts for the lack of fear shown by many insular animals. Darwin noticed that Galápagos birds were so fearless they might land on the barrel of the rifle with which he was trying to shoot them. I have had Galápagos mockingbirds perch on my shoes and nonchalantly pick seeds from the laces as I sat alongside basking iguanas. Recall that one of the predisposing factors for long life is reduced environmental dangers, especially if the reduced danger primarily affects older individuals. Lack of large island predators would clearly reduce one major source of environmental danger.

Human immigrants took advantage of this evolutionary fearlessness to finish off all of these giant island birds that I just mentioned. A few insular giants are still around today, though, including the world's largest bears, living on Kodiak Island, Alaska, and the largest lizard, the Komodo dragon, still eating goats and the occasional tourist on the Lesser Sunda Islands of Indonesia.

On the other end of the island size scale, we have pony-size elephant relatives that Komodo dragons ate before goats and tourists were available. Dwarf elephants also occurred on many Mediterranean islands, including Rhodes, Crete, and Sardinia. California's Channel Islands even had an oxymoronic dwarf mammoth. Madagascar once had dwarf hippopotamuses. In the realm of species with little or perhaps negative charisma, I have trapped dwarf mice on several Micronesian islands. These originally European mice have shrunk to about half their ancestral body size within the last five hundred years after early mariners inadvertently left them behind as they explored the world's islands. Getting back to more charismatic species, in addition to giant lizards and dwarf elephants, Flores, one of the Lesser Sunda Islands, was home to an extinct species of dwarf people, *Homo floriensis*, who at 105 centimeters (three feet, six inches) and twenty-five kilograms (fifty-five pounds) were considerably smaller than any human population today. These dwarf Flores humans—like the elephant birds, the dodos, and the moas—disappeared with the arrival of modern humans. Could we be responsible? Nah, I'm sure it was just a coincidence.

TURTLE ORIGINS

Turtles arose some 220 million years—or as I think of it, several major global extinctions—ago. Turtles are survivors. The first turtles were big but not giant by today's standards. Some later species grew huge, though—many times larger than any of today's giants. When that asteroid collided with earth 66 million years ago, finishing off all nonflying dinosaurs, the flying pterosaurs, the marine plesiosaurs, and all four-legged animals bigger than about thirty kilograms (fifty pounds), turtles were an exception. Eighty percent of all turtle species survived, including *Archelon*, at 460 centimeters (fifteen feet) from head to tail, weighing more than two tons, the biggest of them all. It would be worth knowing how long *Archelon* lived.

Today some 350 species of turtles survive. They live on land, in rivers and lakes, and in the sea. Some species are vegetarian, others are carnivorous, and some specialize in eating jellyfish, whatever that might be called. All, however, bury and then abandon their leathery eggs in terrestrial sand or soil. One of the more bizarre turtle features is that in many species, whether males or females emerge from those eggs depends not on their chromosomes as it does in mammals and birds but on the temperature of the ground in which the eggs were buried. Warmer temperatures typically favor more females; cooler temperatures, more males. For some species, however, males are the Goldilocks sex. In snapping turtles, for instance, the coolest and warmest temperatures produce females, and intermediate (that is, "just right") temperatures produce males. This trait of temperature-dependent sex, once discovered, has been an enormous boon to turtle conservation efforts because captive breeding can be managed to produce a surplus of whichever sex is most needed.

Giant tortoises once occupied many oceanic islands and even some continents across the globe, but today only two groups survive—both on tropical islands. The Galápagos Islands, straddling the equator some thousand kilometers (six hundred miles) west of South America, hosts a number of closely related species of giant tortoises, and the Aldabra atoll, 630 kilometers (four hundred miles) off the east coast of Africa in the Indian Ocean, hosts a single species. These are true island giants in the sense that the closest living relative of the Galápagos tortoises is the Chaco tortoise (*Chelonoidis chilensis*) of Argentina, Paraguay, and Bolivia, which is about the size of a shoe box.

HOW LONG CAN GIANT TORTOISES LIVE?

The long life of giant tortoises is legendary. I mean that in the literal sense. We know that giant tortoises live a long, long time, but exactly how long is still remarkably vague. There is a good reason for this. Unlike the marking and monitoring of bats and birds—in which researchers record initial encounters, mark individuals, and monitor those known individuals until they die or disappear—the marking and monitoring of animals that live as long as tortoises (much longer than humans) would have had to been started a couple of centuries ago, when field naturalists were much more likely to shoot rather than mark and release animals of interest. Also, because considerable acclaim and associated income attends the display of animals of exceptional longevity, age exaggeration in the "oldest animal in the world" game is rampant. Let's examine some of the claims about tortoise longevity and judge them accordingly.

The most extreme claims, while entertaining, can be fairly quickly dismissed for lack of any evidence whatsoever. For instance, in 2019, it was reported that an African spur-thighed tortoise (*Centrochelys sulcata*) named Alagba, a large but not giant species, had died in the royal palace of Oyo state in Nigeria at the remarkable age of 344 years. It should be noted that this remarkable claim was reported by BBC News, one of the most reliable purveyors of unintentionally silly stories about animal and human longevity. Alagba reportedly had great healing powers and—wait for it—attracted visitors from far and wide.

And then there was Adwaita ("one and only" in Sanskrit), a male giant tortoise, possibly from the Aldabra atoll, that lived in the Alipore Zoo, India's oldest zoo, from its opening in 1876 to his death in 2006 when he died at the reported age of 255 years. An alert reader will note that Adwaita lived in the zoo for only 130 years—a very respectable tortoise age—so where did the

255-year age come from? When the zoo obtained Adwaita, it was informed by its then-owner that the original owner was Major-General Robert Clive, who received it following his famous military victory at the Battle of Plassey some 120 years earlier. How much this storied history affected Adwaita's purchase price is not recorded. However, this yarn does indicate that Adwaita was an adult when he arrived at the zoo, so despite the lack of credible evidence that Adwaita lived more than two hundred years, we can be pretty sure that he lived to at least 150 or 160 years, as it takes at least twenty to thirty years for Aldabra tortoises to reach adult size.

There are, in fact, three reasonable (by which I mean somewhat less suspect) longevity records of giant tortoises. The reputedly longest-lived of these is Jonathan, a giant Aldabra tortoise featured prominently on tourist information of the island of St. Helena, where he has apparently lived since 1882. St. Helena, I feel it only fair to report, is one of the most isolated islands on earth, which is why the British government banished Napoleon there to live out the last, sad, lonely years of his life. It is also why St. Helena can use all the tourist dollars it can attract. To wit, Jonathan's great age is widely publicized and his picture is even on their postage stamps and five-pence coin.

Jonathan was apparently a fully mature tortoise when he was brought to St. Helena with three other tortoises. Evidence for this claim is a photo that was reputedly taken in 1886, four years after his putative arrival, which does indeed show a tortoise claimed to be Jonathan, looking fully mature along with four fully mature humans (figure 6.2). Fully mature Aldabra giant tortoises are, according to the St. Helena publicity mill, at least fifty years old—hence the proposed birth year of 1832, making him 189 years young as I write. On the other hand, knowledgeable tortoise biologists consider twenty to thirty years to be the age by which a tortoise of Jonathan's species will be fully mature, as I note above. Both the St. Helena publicity mill and the tortoise biologists could be right. No feature in the trajectory of animal life is as malleable as the length of time it takes to become an adult. That typically relies on energy balance. Individuals that work little and eat much reach adulthood faster than those that work hard for little food reward. In his natural habitat, where thousands of tortoises are competing

Figure 6.2

Jonathan, an Aldabra giant tortoise still living on the island of St. Helena, often erroneously called the world's oldest animal. This photo of Jonathan, the tortoise on the left, has variously been claimed to have been taken in the 1860s, in the 1880s, and in 1902 or "around 1900" (a cropped version showing only Jonathan and the two men standing behind him). As of March 2021, Jonathan was reputedly still alive. Jonathan is clearly very old, but any exact age is pure speculation or perhaps wishful thinking.

for available food, Jonathan might well have taken much longer to become fully mature than he would have if living in an all-you-can-eat zoo.

In any event, Jonathan could be as young as 159 years or even older than 189 years, assuming he is the same tortoise as the one in the photo—which either way makes him quite the codger. Jonathan is not just old: he is aged. Since 2015, he has been blind from cataracts, has lost his sense of smell, and consequently must be hand-fed. Note that long-lived tortoises have many of the same features of aging as humans; they just occur much later.

If you would like to touch the carapace of one of the oldest vertebrates on earth, you should probably hurry to catch one of the weekly flights from Johannesburg to St. Helena.

A second interesting and faintly plausible tortoise longevity record is that of Tu'i Malila, a female radiated tortoise (*Astrochelys radiata*) reputedly given to the royal family of the Tonga Islands by Captain Cook in July 1777 on his third and final voyage of exploration. Tu'i Malila was no giant tortoise; radiated tortoises are more the size of a serving platter. For most of her life, which ended on May 19, 1965 (or some sources say May 16, 1966) at the supposed age of 188 years, she was thought to be a male. Her true sex was discovered only after her death. She also did not have an easy life. Despite being a celebrity tortoise owned by royalty, she was kicked and trampled by horses several times, leaving her badly damaged. Tu'i Malila's major claim to fame aside from her longevity was a group photo taken with the Queen of England, the Duke of Edinburgh, and members of the Tongan royal family during a visit in 1953. When they were introduced to the sprightly, albeit damaged, 176-year-old tortoise, Elizabeth was twenty-seven years old and had been queen for only a year, and Philip was thirty-two years old. Like some other well-known public figures (I'm thinking particularly of Vladimir Lenin, Mao Zedong, and Roy Roger's horse, Trigger), Tu'i Malila's preserved body is still on display in the Tonga Royal Palace for interested tourists.

One key detail about the age of this famous tortoise tweaked my non-sense detector. On his third voyage, Captain Cook, it turns out, stopped at no ports prior to visiting Tonga where radiated tortoises were native. He also never mentioned the tortoise in his journals. Still, it is possible that exotic tortoises were for sale and were bought by crew members at a previous port of call. Whether the royal family would have accepted a gift from a lowly crew member rather than from a captain is psychological speculation beyond my pay grade. A second yarn about Tu'i Malila's origin is that she was obtained by King George Tupou I of Tonga from a passing vessel sometime during his reign, which lasted either from 1820, 1845, or 1875 (depending on how *reign* is defined) to a very final end in 1893. If King George Tupou I purchased or was gifted Tu'i Malila, she was likely much younger but likely at least a century or more old.

The oldest photo of Tu'i Malila I can find is of her being held by Vila Tupou, a younger brother of Queen Salote Tupou III. That photo was likely taken not long before the photo with the new Queen of England. So at least in terms of photographic evidence, the age of Jonathan is much more solidly established.

Finally, there was Harriet, an actual tortoise from the Galápagos Islands kept in recent years at the Australia Zoo and owned by late crocodile hunter Steve Irwin and his wife, Terri. I avoid giving the Latin name of the Galápagos tortoise because DNA analysis has identified some fifteen closely related species living on the Galápagos Islands rather than just one or a few. Zoologists call this a species complex, but for simplicity, Harriet is just a "Galápagos tortoise."

To keep alive the theme of celebrity ownership, Harriet was reputedly captured from the wild in 1835 by none other than Charles Darwin himself. How Harriet reputedly made the journey from the Galápagos Islands to a zoo in Beerwah, Australia, may be more circuitous than the voyage of the *Beagle* itself.

One thing that is noticeable about all of these longevity claims is that the chelonian chain of custody always goes murky in one or several places during the turtles' lives. Harriet's murky chain of custody sets a new standard, though. The oft-repeated story of Harriet is that Darwin collected three juvenile tortoises during his visit to the Galápagos and took them back to England to keep as pets. At some point, he gave them to John Wickham, a former shipmate. Wickham upon his retirement from the Royal Navy migrated to Brisbane, Australia, in 1841, taking his three pets (now named Tom, Dick, and Harry) with him. They lived with him until 1860, when he donated the three, now fully mature, tortoises to the Brisbane Botanical Gardens zoo. From there, Harriet was purchased by famous naturalist and zoologist David Fleay in 1952, when the Botanical Gardens shuttered its zoo. Fleay, being an excellent naturalist, is the one who discovered that Harry should really be named Harriet. Turtle sex identification clearly takes a trained naturalist's eye. Fleay kept Harriet with him until 1987, when she was transferred to the Queensland Reptile and Fauna Park, then owned by Bob and Lyn Irwin, Steve's parents, and later renamed Australia Zoo when Steve and Terri took it over. So are we clear? It was Darwin to Wickham to Brisbane Botanical to Fleay to Irwin. In her later years, Harriet (aka Darwin's tortoise) became quite the attraction before her death in 2006. Assuming that all of that history is true, she would probably have been hatched in 1834 or a few years earlier and so lived to be at least 172 years old.

It may seem churlish to question this colorful history, but there are a few real problems with it. For one thing, it is contradicted by the contemporary account of Darwin's servant and companion aboard the *Beagle*, Syms Covington. According to Covington, Darwin brought back only one juvenile tortoise, and Covington himself brought back another. These would not likely have been pets. Darwin was not a "pet" person. He was more a "study specimen" person. That they were not pets is also pretty clear from an inquiry about them Darwin received many years later, at which time he didn't remember bringing back any tortoises at all. Also, census documents record Wickham as living in Australia at the time he was reportedly receiving them from Darwin in England. Darwin himself did not recall meeting Wickham after the voyage was over until a *Beagle* reunion some twenty years later. Perhaps the final and most emphatic nail in the coffin of this story of Darwiniana is a recent discovery by Colin McCarthy, a collections manager at the Natural History Museum in London. McCarthy, in preparing for the 2009 Darwin Bicentennial Exhibit, found in the museum basement, in Zoology Dry Storeroom No. 1, two juvenile tortoise skeletons, both registered on August 13, 1837—ten months after the return of the *Beagle*—with labels that read "presented by Charles Darwin Esq."

While Harriet never seems to have been Darwin's tortoise, that does not mean that she may not have been pretty ancient, even as ancient as claimed. Science journalist Paul Chambers, whose research I've used for much of this account, tracked down an unpublished analysis of Harriet's DNA, collected after her death. That analysis indicated that Harriet came from a Galápagos island Darwin never visited, Santa Cruz Island, known as Indefatigable in Darwin's day. However, it also showed something else.

In the middle 1800s, thanks largely to the excellent maps made during the *Beagle* voyage, the Galápagos Islands became a popular port of call for whalers plying the tropical waters where sperm whales were abundant. The 1849 California gold rush also brought a flurry of passenger ships sailing "round the Horn" (Cape Horn) carrying '49ers in search of fortune. Tortoises were plundered during virtually every ship's visit, as tortoises with their slow metabolism could be stored alive in a hold for up to a year without food or water, and their meat by all accounts was scrumptious. In the decades after the *Beagle*'s visit, thanks to plundering by passing mariners, tortoise populations collapsed. From a prehuman population estimated at a quarter million, by the mid-twentieth century there were only about three thousand giant Galápagos tortoises left. Conservation efforts have now reversed that trend, but having passed through a severe population bottleneck, the survivors' DNA is still recognizable but noticeably altered from what it was previously. Harriet's DNA turns out to be similar enough to existing Santa Cruz tortoises to be convincing that she came from that island but different enough to suggest she came from that island a long time ago—maybe even as long ago as when Darwin visited the archipelago.

So as I said at the beginning, tortoises can live a very long time but exactly how long remains a mystery. Somewhere between 150 and two hundred years seems like a reasonable guess though.

Notice that I haven't estimated longevity quotient for any of these tortoises. That has nothing to do with the fact that establishing their true age within a few decades is problematic. Instead, it has to do with their heavy shell. LQ was established from the longevity of nonshelled mammals generally kept in zoos or as pets, and it depends on weight. To a crude extent, the weight of most vertebrates is related to their overall body volume. However, the shells of turtles and tortoises make them considerably heavier than nonshelled animals of the same volume, so making comparisons with other species can be deceptive.

One other exceptionally long-lived island species, this one without a shell, deserves mentioning here.

TUATARAS (*SPHENODON PUNCTATUS*)

Another island reptile is the tuatara, a rare and very strange animal even by island standards. Many tuatara species lived throughout the southern continents during the heyday of the dinosaurs, but today only a single species survives on thirty-two small islands off the coast of New Zealand, which is itself composed of two larger islands. Tuataras hold an iconic place in indigenous

Māori culture, where they are considered a special treasure (*taonga*), guardians of special places. They were represented on the now defunct New Zealand five-cent coin, and New Zealand's only professional baseball team is the Auckland Tuataras.

Tuataras look like lizards but aren't (figure 6.3). In fact, they diverged from lizards and snakes some 250 million years ago—before dinosaurs existed. Compared with their long-extinct ancestors, they are island dwarfs rather than giants. Adults are only some sixty-centimeters (two-feet) long and one kilogram (2.2 pounds) in weight. Adults are also strictly nocturnal and ground-based, whereas juveniles are active during the day and spend most time in trees, probably to avoid the seabird predators with which they share their islands. They have excellent low-light vision as might be expected from a nocturnal predator. They may even see colors in low light. This has nothing to do with their third eye, complete with lens and retina, on top

Figure 6.3
The tuatara is not a lizard. It is an ancient reptile that diverged from snakes and lizards 250 million years ago. The name means "spines on the back" in the Māori language. The slowest-growing reptile is one of the longest-lived species as well. This is an 1886 engraving.
Source: Photo by George Bernard/Science Photo Library.

of their head, which is visible to our normal eyes for the first few months of juvenile life, after which it is covered by scales. Their single row of lower teeth fits nicely between the two rows of teeth in their upper jaw. As these are permanent teeth, grinding up abrasive food will wear them down with age. Young tuataras often dine on crunchy items such as crickets and beetles, but as they age and their teeth get smaller and duller, eventually becoming little more than nubs, they switch to earthworms, slugs, and insect larvae, which they can gum to death.

Tuataras are not in a hurry. They have evolved at a glacial pace for 250 million years, and they live the same way. They are the slowest-growing of reptiles, in the wild taking thirty-five years or more to reach final adult size. They spend eight months to a year inside the egg and first reproduce as teenagers. Females in the wild lay eggs only about every four years. This may be because the live in a cold, foggy climate and have the lowest preferred body temperature (16°C to 21°C [61°F to 70°F]) of any reptile. Like tortoises, tuatara sex is determined by temperature. Opposite to most tortoises, though, low temperatures produce mostly females, and higher temperatures produce mostly males. There is concern among some tuatara conservationists that global warming may lead to wild populations ultimately producing only males.

How long can these slow-growing, cold-living reptiles survive? There have been rumors of hundred-year-old, even two-hundred-year-old tuataras for years without any solid evidence. Then I came across a 2009 CNN report that a 110-year-old male tuatara named Henry, living in New Zealand's Southland Museum and Art Gallery (SMAG), had hooked up with youthful eighty-year-old Mildred to produce eleven healthy new hatchlings. The details were as intriguing as the main story was suspicious to me. Henry, it seems, had been a cranky old fellow with no interest in females for at least the previous thirty-five years, until veterinarians discovered a tumor in Henry's so-called nether regions. I would be cranky too. After its removal, his attitude improved, his libido returned, and the next thing you know, he and Mildred had produced clutch of young ones. To say I was dubious is an understatement, but to learn more, I contacted SMAG's curator and tuatara breeder par excellence, Lindsay Hazley.

Hazley calls himself the *kaitiaki* (guardian) of the SMAG tuatara colony, which he has managed for nearly fifty years. During that time, he has personally worked out most of the husbandry issues needed to keep tuataras healthy and willing to breed in captivity. In fact, he has distributed about seventy captive-born tuataras to zoos throughout New Zealand to help advocate for continued protection of the species.

My first question was, "How do you know Henry's age?" As I suspected from the round number (110 years), this was an estimate, but it was an estimate based on sound, long-term observations. Henry has lived at SMAG since he was captured as a full-size adult in 1970. Hazley has in his colony males like Henry that he has watched grow from egg to adult for over thirty years. Based on these growth rates, he estimates that his captive-born males will require another twenty to thirty years to reach Henry's size at arrival. Given the SMAG colony is kept at 15°C to 17°C (59°F to 63°F), which is 5°C to 6°C (9°F to 11°F) warmer than the natural habitat where Henry grew up, and knowing that colony-reared animals get more abundant food for less work than wild tuataras, Hazley calculates that Henry was around sixty years old when he arrived.[2] So add fifty years at SMAG to sixty years of wild and free life on cold, foggy, windy Stephens Island, where Henry grew up, and there you have it. Yes, it is an estimate but a reasonable one, it seems to me. Using Henry's estimated age, tuataras have a longevity quotient of about 10.3, a bit longer-lived for its size than Brandt's bat.

WHAT CAN WE LEARN ABOUT THE BIOLOGY OF LONGEVITY FROM LONG-LIVED TORTOISES AND TUATARAS?

Understanding the longevity of tortoise and other reptiles must start with their ectothermy. That is, they do not produce their own heat: their body temperature is dictated by the temperature of their environment. In the Galápagos Islands, the annual average daily temperature is 24°C (75°F), which would be a good approximation of the average body temperature of a Galápagos tortoise. Tuataras in the wild live at an annual average 9°C to 10°C (48°F to 50°F), which would be their average body temperature.

Rather than undergoing the tremendous energetic cost of maintaining a temperature many degrees higher than their surroundings—like, say, the Galápagos hawk (*Buteo galapagoensis*), with its 41°C (106°F) body temperature—long-lived reptiles have adopted an energy-saving strategy by slowing everything down. Giant tortoises and tuataras move slowly, grow slowly, and reach life's milestones slowly. Darwin calculated tortoises could travel up to four miles *per day*, not *per hour* like a person in a brisk walk. In well-fed captivity, it takes them at least twenty to thirty years to reach tortoise puberty, and in the wild it takes forty years or more. Their heart beats slowly—only once every ten seconds. They breathe slowly. Their cells divide slowly. We don't know how fast they think, but you probably would not want to play chess with one.

Due to this energy-saving strategy, reptiles generally and tortoises and tuataras, in particular, need to eat only a fraction as much as a similar-size mammal. As with birds and mammals, larger reptiles have lower metabolic rates than smaller reptiles, which is why giant tortoises could be stored in a ship's hold without food or water for a year or more.

The general sluggishness of their life makes exceptional tortoise and tuatara longevity very different from that of birds or bats. Bird and bat longevity is exceptional relative to their body size, which in endothermic animals means relative to the rate at which they process energy. In absolute number of years lived, they do not approach humans, much less tortoises. Remember, the longest-lived zoo bird ever lived to age eighty-three (Major Mitchell's cockatoo), which would not be exceptional for a human. However, I am certain that we have much to learn about extending human health and longevity from birds and bats because of the rate at which they have lived those years. During his life, Major Mitchell's cockatoo processed more than nine times the amount of energy, with all the damaging chemical byproducts that implies, than the longest-lived humans do. Because they can process more energy, much faster, than humans and because energy and temperature speed up both beneficial chemical reactions (like turning food and oxygen into a beating heart and a thinking brain) but also accelerate deleterious chemical reactions (like oxygen radical production, the production of browning products, and a number other damaging products), we have a great deal to learn from them about how to maintain health in the face of the damaging fire of life.

Giant tortoises (even though they can certainly live at least 150 years and probably several decades longer) and tuataras (assuming they live at least a century) process only a fraction of the energy during their long lives compared to humans. Their hearts have beat far fewer times, and they have drawn far fewer breaths than an average human. Their dramatically slower metabolism and lower body temperature mean fewer oxygen radicals will be produced over time and protein modification into browning products will proceed more slowly. In fact, as aging is almost entirely due to damaging by-products of cellular chemistry and since almost all chemical reactions proceed slower at lower temperatures, it would be biologically astonishing if giant tortoises and dwarf tuataras were not exceptionally long-lived.

This does not necessarily mean, however, that we have nothing at all to learn about long life from these reptiles. It could indeed be that their longevity is due to little more than living life in the very slow lane, but there could conceivably be more to it as well. Particularly given their large size, meaning their large number of cells any of which could potentially turn cancerous, we might expect exceptional cellular defenses against cancer. Early analyses of their genomes, in fact, suggest this may be the case.

The great scientific achievement of the twenty-first century so far has been our remarkable ability to sequence genomes. The linear sequence of the four DNA "letters," called nucleotides, that compromises a genome, is a universal code that determines what genes the genome contains, how and when those genes are turned on and off, and what cellular tasks those genes perform. The first human genome sequence, all three billion nucleotides— about the number of letters in a thousand King James bibles—took a decade and a billion dollars to complete by 2003. Today, not even twenty years later, it takes a few hours and costs less than dinner at a nice restaurant to sequence an animal's genome. So something that can be quickly and cheaply done these days is to sequence the genomes of long-lived species. Okay, I am oversimplifying a bit. Deciphering what biological information the code reveals still takes considerable work, but we are getting better and faster at that too.

Genome sequences can help us locate promising areas for understanding longevity. For instance, recent work on the genome sequences of long-lived tortoises suggests that DNA damage repair and cellular resistance to cancerous transformation are major processes beyond a slow metabolism that may contribute to their exceptional longevity. However, genome sequences can only suggest—they cannot determine. They are a start in understanding the biology of longevity, but only a start.

To understand whether knowledge from particular long-lived species can be used to improve or extend human health, we first need to learn whether the species in question combats an aging process or processes more successfully than humans already do, and if so, then learn how they do it. It is a tall order that requires a major research effort far beyond that of merely sequencing genomes. Alas, nearly all of our research effort currently goes toward the deep study of species that we already know are pitifully unsuccessful at combating aging processes relative to humans. We consider it a triumph when we make them a little less unsuccessful. My hope is that this book may inspire the deep study of the most promising species from Methuselah's Zoo, because that, in my humble opinion, is the key to developing medications leading to longer, healthier human lives.

7 QUEEN FOR A LIFETIME

Imagine that one day you discover an exceptional woman from an otherwise normal family. Her normal family has produced many generations of normal relatives. The exceptional thing about this woman is that unlike her relatives—all of whom have lived the normal seventy to ninety years or so—she has been alive for more than two thousand years. She has lived through Greece's Golden Age, the rise and fall of ancient Rome, the plague-ravaged Middle Ages, the blossoming Renaissance, the Age of Enlightenment and rise of science, the astonishing wealth and air pollution that accompanied the Industrial Revolution, and the spread of technology, and today, in the Digital Age, she is still in good health. Would learning the special traits that allowed this woman to survive so many times longer than anyone else be of interest to you? It certainly would be to me. Remarkably enough, nature has provided us with examples of just such a scenario—not in humans, of course, but in insects.

Appropriately enough, we call these extraordinary survivors "queens." Queens live in robust health an extraordinarily long time and as much as thirty times as long as their nonroyal relatives. In human terms, thirty normal lifetimes would push the birth date of someone alive today back to around the time of Aristotle. The best examples of such queens can be found, prosaically enough, in some of our most common and commonly despised insects—ants and termites. A distinctive feature of both ants and termites, in addition to the extraordinary longevity of their queens, is the large, complex societies in which they live. These two things—extraordinary queen longevity and complex societies—turn out to be related.

Ant and termite queens and societies have certain important features in common. One of these features is that they were first described in the days when scientists didn't think much about the anthropomorphic implications of their terminology. So in addition to queens, we find ant and termite social roles called castes and specific castes called workers, drones, kings, nurses, soldiers, and even slaves. These days, there are technical, albeit obscure, names for these roles (such as macrergates, pseudergates, and dinergates) just as there are obscure technical names for their body parts (mesosoma and metasoma instead of thorax and abdomen, respectively). But rather than wandering through a thicket of obscure scientific jargon, let's just agree that there are no intended anthropomorphic implications of these commonly used entomological terms and carry on using common language.

Ant and termite colonies consist of usually one but sometimes several multigenerational families. These families may be the size of a tiny village or a large city, depending on the species and environment. The families live in well-protected nests, in extensive systems of excavated tunnels and chambers underground, in above-ground citadels, or in wood. Because the families live in the dark, they communicate with chemical signals. All food and labor to support the family's needs are provided by workers. Adults do not typically reproduce. Only royalty is allowed that privilege.

Ant and termite societies differ in interesting ways too.

ANTS AND THEIR SOCIETIES

If it seems to you that ants are everywhere, you are right. They are native to all continents except Antarctica. They are found in stupefying abundance in deserts, plains, and forests as well as in parks, playgrounds, and pantries. In the tropics, a football-field-size piece of ground may contain some 5 million ants. Some species are voracious predators, consuming vast quantities of spiders, other insects, and diverse small soil animals such as millipedes, centipedes, worms, and termites. Others drink nectar or imbibe the excretions of aphids. Some even eat specialized plant parts produced just for them. Still others farm fungus gardens inside their nests, having invented agriculture at least 50 million years before humans. In deserts, ants store huge larders of seeds. Because they build large underground burrow systems, ants churn and turn as much soil as earthworms in northern climates and much more than earthworms in the tropics. Because they carry the corpses of their animal or vegetable prey back to decay in their burrow systems, ants provide more than a little soil fertilizer.

Like any other animal that competes with humans for food and space, ants are considered pests, and a multibillion-dollar industry is focused on their extermination. One famous ant elimination effort was the DDT carpet-bombing of vast stretches of the southeastern United States in a vain attempt to exterminate imported red fire ants, themselves responsible for massive damage to buildings, agriculture, and livestock. Ironically, that event—which did more damage to pets and non-ant wildlife than to the fire ants—caught the attention of writer Rachel Carson, whose 1962 book *Silent Spring* about the dangers of indiscriminate pesticide use is often said to have launched the environmental movement.

Ant societies are female societies. The queen is a female, of course, but so are all of the workers. Workers earn their name. They do all the foraging. They transport food back to the nest. They excavate and maintain new nest chambers and tunnels as the colony expands. They care lavishly for the brood, a collective term for all immatures whether in egg, larval, or pupal stage. They feed the queen, one another, and the brood with food that has been processed in their guts. They move the brood around as necessary, tend food stores, keep the colony clean, and provide suicidal colony defense as required.

Ants hatch from eggs into tiny wormlike, legless, eyeless, antenna-less, completely helpless larvae, which need tending as much as any human baby. Workers feed the larvae regurgitated liquids early on and solid food when they are ready for it. They lick the brood and remove their waste—the ant equivalent of changing diapers. They also assist them with molting. Larvae are eating machines. They grow and molt—usually four times—into successively larger but still helpless larvae, until they eventually spin a cocoon, becoming immobile pupae, and emerge a few weeks later in final adult form and size. As adults, ants, like other insects, do not molt, so they are capable of very limited bodily repair if and when they are damaged.

The fifteen thousand or so ant species display many variations on one common colony life history, although there are special cases. That typical history begins when a new colony is founded by a single winged female who flies away from her natal colony among a squadron of other winged females and winged males. During this flight, the only flight that she will ever make, she mates. When she lands, she breaks off her wings and digs the nest chamber where she will spend the rest of her life. No such future awaits the male consorts with whom she flies. Their single job is now accomplished, and they soon die.

The now wingless queen shortly begins laying eggs. As those eggs hatch into larvae, she will feed them regurgitated food from energy stored as fat or as no-longer-necessary flight muscle. Once those first generation of eggs reach adulthood as workers one to two months later, they take over virtually all colony tasks, including foraging and feeding the queen and larvae emerging from the next generations of eggs. The labor of colony founding has reduced the queen to half her original weight, so the workers will be particularly attentive in feeding her. Once this stage is reached, the queen becomes little more than an exquisitely pampered egg-laying machine for the rest of her life.

Males, as I note, are transitory and useless except for mating. Workers feed and tend them like any other larvae as they develop, but their only contribution to colony life will be to pass sperm to incipient queens during their own single mating flight. The life of a male ant may be short, but the life of their sperm is not. Queens store sperm in specialized organs where it will be dolloped out to fertilize eggs for the rest of the queen's long life.

If males are rare compared to females, except for certain times of the year when the colony swells with them, one obvious question is how is ant sex determined to produce such a lopsided ratio of sexes? The answer is a biological trick shared with bees, wasps, and a few other insect groups. Females develop only from fertilized eggs. Unfertilized eggs, instead of being stillborn as they would be in most species, develop into males. As eggs pass down the queen's birth canal day after day, week after week, year after year, she can either release a bit of her stored sperm to fertilize the passing eggs or not. If she does, she lays a female egg. If not, the egg becomes a male. Colony sex ratio is the queen's prerogative.

This unusual system of sex determination has some strange genetic consequences from a human perspective. It means that males carry only half the genetic material of females and that all of a male's genes are inherited from his mother. Females, on the other hand, have half of their mothers' genes and all of their short-lived fathers'. Consequently, sisters (that is, the colony workers) turn out to share three-quarters of their genes with one another—half of their genes through their mother, plus only a quarter of their genes through their short-lived, largely useless brothers. Sisters are closer relatives to one another than they are to their parents, in other words. Some researchers have attributed the evolution of their complex societies to this unusual genetic system.

Another question—the billion-dollar question for us—is what determines whether a female egg develops into a relatively short-lived worker or a long-lived queen and what, if anything, might we be able to learn about lengthening healthy human life from understanding what makes a queen. But before we get to that, let's compare ant to termite societies because termite queens are also exceptionally long-lived. Comparing these two groups may tell us something general about the evolution of exceptionally long insect life.

TERMITES AND THEIR SOCIETIES

Emphasizing their superficial similarities, termites are sometimes called "white ants," although in actuality they are not closely related to ants at all. They *are* closely related to cockroaches, another of our favorite insects. In fact, modern biologists consider termites a subgroup of cockroaches. Unlike ants, termites specialize in eating dead material at many stages of decomposition, a habit called *detritivory*, although they consume living plants as well. As such, they are premier recyclers. However, if they happen to be recycling the wood in your house or the books in your library, as they did to me when I lived in the tropics, you may not consider this such a laudable public service. Some termite species, like some ant species, farm living fungus inside their nests. Without much prompting, termite biologists will legitimately brag that termites invented agriculture even earlier than ants. They were farming when dinosaurs still roamed the earth.

Like ants, termites are seemingly ubiquitous, also being found on all continents except Antarctica. In the tropics, their abundance is particularly stupefying. Together, tropical ants and termites are so numerous that some good-size mammals—such as anteaters, aardwolves, armadillos, pangolins, and numbats—eat little else. Both ants and termites can be exquisite architects—ants usually building underground labyrinths and termites often constructing spectacular above-ground structures.

At first glance, ant and termite societies are strikingly similar, too, consisting of numerous variations on the same basic theme. Termite colonies are also founded by winged adults, who break off their wings soon after alighting from a single nuptial flight. They also excavate a nest and never leave that nest again. Typical termite colonies are also multigenerational families descended from eggs laid by a single founding queen. Almost all of those eggs develop into wingless sterile workers of various sorts who gather food and perform all necessary labor of the colony including feeding the queen, her brood, and one another.

For all this similarity, there are some differences worth noting, too. Unlike ants, termite queens do not mate during their single nuptial flight. That flight is all about finding a life partner. As partners, a winged male and female will land together and break off their wings together. They dig a nest into soil or dead wood together, and only after they have done all these preliminaries, do they finally mate. They continue to live together and mate until death do them part. Compared to ants, termites are romantics. Thus, there are termite kings as well as queens.

Unlike ants, termites do not undergo a complete metamorphosis from helpless, legless, antenna-less larvae that pupate into very different-looking adults. Termite larvae have legs, and although they are helpless upon hatching, as they grow and molt, they are called nymphs, which resemble smaller adults. Older nymphs even perform some colony functions. The developmental system of termites is much more flexible and complicated than that of ants. Some even have an "arrested development" stage where they cease further molting but can resume it, developing into new reproductive adults if the colony need arises.

Also unlike ants, both sexes, not just females, fill all worker roles and are produced in approximately equal numbers. Termite sex is determined as it is in mammals. Females inherit two X chromosomes, one from each parent. Males inherit an X from their mother and a Y from their father.

ANT AND TERMITE LONGEVITY

Recall from chapter 2 that adult insects who spend their lives out-and-about, so to speak—that is, they fly, forage, and mate out in the open like grasshoppers, butterflies, and actual flies—are short-lived. As adults, they typically live weeks or months—at the longest, perhaps a year. This is because their small size makes them vulnerable to many sources of environmental danger and also because once insects have undergone their final molt into adulthood, they have virtually no ability to repair or replace damaged parts, and damage is inevitable. They never developed nor needed a longevity kit.

For ants and termites, life is very different. We need to be careful here because like Alagba, the spur-thighed tortoise, or any number of human longevity myths, there are some unverified longevity whoppers out there, particularly about termites. Let's face it, individual insects—particularly those that live underground in colonies that may number thousands to millions of individuals—are not easy to monitor for weeks or months much less years or decades. Having said that, we do know from laboratory colonies that ant and termite queens can live as long as several decades as nonmolting adults. The oldest validated laboratory age a for termite queen is twenty-one years—twenty times longer than any known above-ground insect—and several of that age were still alive at last report, so they could live even longer. These twenty-one-year-old queens are giant northern termites (*Mastotermes darwiniensis*) of Australia, a species that got its name because winged queens and kings are, yes, big—averaging thirty-five millimeters (1.4 inches) in length. While we are on the topic, northern giant termite kings live every bit as long as queens. This species may not be all that exceptional for a termite. Even though we have limited information on only a handful of the four thousand or so termite species, generally those of particular economic importance, we do know of at least five other species in which queens live into their teens and possibly longer.

Ant queens appear even more impressive. The longest-lived with an unassailable birth certificate, so to speak, is a twenty-nine-year-old queen of the common black ant (*Lasius niger*), one of the most common European ant species. How unusual is that age for an ant queen? Like termites, we have no idea because we know precious little about precious few ant species. Common black ants happen to be particularly well known because they are easy to maintain in captivity and so are a favorite of ant hobbyists. However, we know of at least six other ant species that have verified queen ages of more than twenty years in the house or laboratory and one seed-eating harvester ant (*Pogonomyrmex owyheei*) even has reasonably believable credentials for living at least thirty years in the wild.[1] Ant queen longevity is truly extraordinary among insects.

Both ants and termites break one of the cardinal longevity rules we have encountered so far. Specifically, the rule is that long life requires slow development and a low reproductive rate, as we've seen in reptiles, birds and mammals. Remember that among birds, the long-lived Manx shearwater takes six years to reach adulthood and lays one egg per year compared to the short-lived turkey, which matures in a year and lays a dozen eggs at a time. And among mammals, long-lived bats take one to two years to reach maturity and have a single pup per year compared with same-size, short-lived mice that are adults in two months and can produce litters of six or so pups every six weeks. Also, don't forget that long-lived tortoises require decades to reach adulthood.

Compare this with long-lived ant and termite queens that become adults within a month or two, continuously produce as much as an egg a minute for the rest of their lives, and yet are the longest-lived insects known. Along the same lines, they also seem to violate the loose but general pattern that longer life is related to lower metabolic rate. Egg production in queens requires much more energy than that expended by workers tending larvae, for instance, but despite that they live many times as long as workers.

However, a general pattern that ants and termites follow precisely is that species occupying environmental niches that protect them from external hazards live a long time. Ants and termite queens are protected from external hazards by living underground, inside their exquisitely constructed citadels, or inside wood and having thousands of potentially suicidal workers to feed and protect them. I should note here that ant and termite workers can be pretty long-lived themselves, living up to three or four years in some species, particularly if they fill colony roles that never require them to leave the nest. Workers that regularly leave the nest to either forage or fight are as short-lived as any other insect. Further supporting this pattern is the fact that some ant species—particularly those with more than a single queen—do not construct such elaborate protected nests, and the entire colony will periodically move above ground to a new location as environmental conditions change. Queens in these colonies are not as long-lived, even in the laboratory, as queens who spend their entire lives underground. A final fact supporting this pattern is that honeybee queens—who live in a social system nearly identical to ants, except that hives tend to be above ground thus less well-protected—are much shorter-lived than ants, although still considerably longer-lived than honeybee workers. Among professional bee researchers, five years is considered to be about the longest that a honeybee queen can survive.

I will point out in passing that there is one account published in a Russian beekeepers' journal in the early 1960s of a honeybee queen living eight years.[2] That account is so inconsistent with thousands of other observations by professional beekeepers that it is widely discounted in the beekeeper community. However, it is consistent with grossly exaggerated human longevity claims also coming out of Russia in the 1960s (more about that later).

Another pattern that ants and termites often follow is the size rule: larger size classes live longer than smaller ones. One thing I haven't mentioned yet is that workers within a single colony can vary as much as fifteen- to twenty-fold in size, but the queen is even larger—sometimes much larger—than any of them. Getting back to the imaginary example of the exceptional woman who lived thirty times as long as any of her relatives, to keep the analogy going she would also have been exceptional in one other aspect. She would have been enormous, maybe three meters (ten feet) or more tall.

Imported red fire ants (*Solenopsis invicta*), because of their economic importance, are probably the best-known ant species and one of the few in which we know details about worker longevity (figure 7.1). Walter Tschinkel (or Dr. Fire Ant, as I think of him) is a professor at Florida State University who has spent an illustrious career learning pretty much everything there is to know about fire ants. From data lovingly and laboriously gathered by his

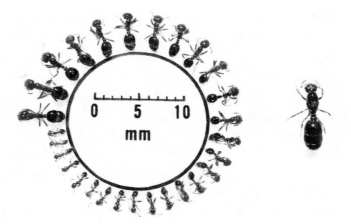

Figure 7.1
Size variation in imported red fire ant workers (circle) and queen (right). Larger workers live longer than smaller workers, but the queen can live twenty-five times as long as even the largest worker.
Source: Photo courtesy of S. D. Porter.

students on the three size classes of fire ant workers, we learned that small workers lived on average fifty-one days, medium workers eighty days, and large workers 121 days in the laboratory. Fire ant queens are much larger than any of the workers, as I've noted, and although they are not particularly long-lived among ant species, they still live up to eight years or some twenty-five to sixty times as long as the workers who provide for them.

So protection of adults from environmental hazards appears to afford evolution an opportunity to provide many species with a longevity kit—the physiological capability to live a very long time—and evolution often takes advantage of that opportunity. An intriguing question is whether that longevity kit always contains the same assortment of tools.

HOW DO THEY DO IT?

Ant and termite queens and workers do not differ in their genetic inheritance. That is, in most species of ants or termites, any female egg has the potential to develop into either worker or queen. So what makes the difference between an egg that becomes a short-lived worker or a long-lived queen?

For many years, ant biologists assumed that incipient royalty were fed special, almost magical, nutrients as larvae and that these special nutrients accounted for their long life. This was not a crazy idea given that close observation showed that workers did feed larvae some different regurgitated substances if they were rearing a queen rather than rearing another worker. I should note that this was also true for honeybees, in which this special substance had a special name—royal jelly. Royal jelly is still marketed in health food stores and on the internet as having remarkable health-enhancing properties, almost a miracle of anti-aging power. Skeptics like me used to say these claims might be accurate—if you are a honeybee. But now we know that they aren't true even if you *are* a honeybee. Laboratory studies have clearly shown that honeybee larvae fed a wide variety of diets can emerge as queens, provided they are fed enough of it.[3] Quantity turns out to be more important than quality.

Readers may have noticed that I have not mentioned what determines whether a termite nymph develops into a queen (or king) instead of a soldier or worker. The reason is that we know a great deal less about that than we know about this development in ants (or bees). What we do know suggests that termite and ant caste determination work reasonably similarly to one another. So I will concentrate on the group where the most reliable information is available.

In ants, the prosaic truth is that larval size determines what caste—queen or otherwise—larvae ultimately become, and what determines larval size is complicated. It is biology, after all. Biology is complicated. If it were simple, it would be physics.

Nutrition certainly plays a role, but as with bees, quantity, as much as the quality of nutrition, matters. Hormones also count. Queens play a role here as they can produce eggs with more or less of a hormone called *juvenile hormone* that, along with nutrition, determines how large the developing ant will become. Queens also produce odors—that is, pheromones—that suppress growth and ovary development in the larvae. Colony temperature also plays a role. As I said, it's complicated and not terribly relevant to human longevity.

There is another level of explanation, though, that may be relevant. Larvae destined to become queens versus workers may share the same genes, but that doesn't mean that the same genes are active. All the cells in your body have the same genes, too, but some of those cells are brain cells, and others are muscle cells, liver cells, blood cells, and so on. What determines the type of cell they become and what they do once they have assumed their final form and function is which genes are turned on (that is, active) and which are turned off. This is similar for ant queens versus workers. Can we learn something about longevity genes from comparing gene activity in ant queens versus workers?

Due to their active metabolism, we might guess that queens should have much more active cellular damage prevention and repair genes turned on than workers. That guess would be correct. A veritable swarm of protection and repair genes are turned on in queens relative to workers. One ant gene appears to be of particular interest. Daniel Kronauer of Rockefeller University and his colleagues discovered a single insulin-like gene, and only that gene was highly active in ant queens but not in workers in seven ant species.[4] That gene is prosaically named ILP-2, for "insulin-like peptide 2." Insulin itself and insulin-like genes are involved in metabolism, growth, and reproduction, and they hold a special place in aging research. The first gene discovered to affect longevity in multiple short-lived laboratory species (like worms, fruit flies, and mice) was an insulin-like gene. Of two insulin-like genes found in ants, ILP-2 most resembles human insulin. Here is the evolutionary puzzle, though. Increased longevity in laboratory species is accomplished by reducing the activity of insulin-like genes. Long-lived ant queens have *increased* insulin-like activity relative to shorter-lived workers, which is the exact opposite. So quite possibly, ant queens achieve their great longevity due to the activity of a very special insulin-like gene, meaning that we may have something novel and important to learn about achieving long life from studying that gene. Those studies are in their infancy, but I look forward to what they will teach us in the not-too-distant future.

As we have seen, among the insects, flight does not predispose species to long life. The key predisposing factor seems to be a protected, underground life style. It turns out that insects are not the only group for which this is true, as we see next.

8 TUNNELS AND CAVES

One of my academic friends calls it "a small penis or large thumb with teeth." It lives underground in a colony of several dozen to several hundred relatives in a complex series of tunnels and chambers that may meander for more than a kilometer. It never leaves its underground warren and never sees the sun because its food—large tubers, which is where plant root systems store their energy—is also found underground. Like termites, there is a single queen who does all of the colony reproduction, and one or perhaps several kings who ensure that the queen is always inseminated so that she can pump out pups as fast as possible. Sterile workers, all close relatives of both sexes, make up the rest of the colony. Some workers excavate, repair, and clean the burrow system; others provide food in the form of nutritious feces to the queen and developing young. Some carry the young when necessary. The largest workers serve as soldiers, protecting the colony from attack by enemies foreign or domestic—that is, intruders of its own species or predators such as snakes.

It is a mammal, but unlike other mammals it has little ability to regulate its body temperature and so assumes a body temperature close to that of its surroundings. Because it does not spend an inordinate amount of energy to heat itself, its energy needs—that is, its metabolism—is slow relative to other mammals of similar size. From this description (lives in an environmentally stable habitat, is well-protected from external dangers, has a low metabolic rate), if there are reliable patterns in nature, you might predict that this creature (at least its queens but maybe even the workers) should be long-lived. You would be right. It is the only species in my top twenty-five longest-lived nonhuman mammals (based on longevity quotient) that is not a bat. The animal I'm describing is the mouse-size naked mole-rat (NMR, for short), either the ugliest or the cutest of mammals, depending on one's taste (figure 8.1). I subscribe to the former.

Naked mole-rats (*Heterocephalus glaber*) are neither moles nor rats. They are in a small and unique family of burrowing rodents that are widely distributed in arid parts of sub-Saharan Africa. They are about as distantly related to rats as rodents can be, their ancestors having diverged at around the time that the nonavian dinosaurs and pterosaurs disappeared. They are even more distantly related to moles, which are not rodents at all. The only similarity to moles is that they both live underground and dispose of excavated dirt from burrow construction in characteristic, above-ground "molehills."

NMRs came to biologists' attention in the late 1970s, when one of the burning evolutionary questions was why vertebrates had never evolved eusociality. Eusociality is the social system we saw in ants and termites—multigenerational families living together, extensive division of labor, with

Figure 8.1

A thirty-seven-year-old naked mole-rat in all its charismatic glory. About the same size as a mouse, naked mole-rats live up to seventeen years in the wild but more than twice as long in the laboratory. According to Rochelle Buffenstein, this same animal was still alive at thirty-nine years of age.

Source: Photo courtesy of Rochelle Buffenstein.

reproductive queens and sterile workers performing a variety of tasks to support the queen's reproductive efforts. Insects had clearly evolved eusociality multiple times. Why not vertebrates? The evolution of eusociality was a burning question because it challenged the dogma that evolution favored only traits promoting individual reproduction. In eusocial groups, the overwhelming majority of individuals never reproduce. Instead, they helped others reproduce, an act of seemingly extreme evolutionary altruism.

Then in 1981, the same year I received my PhD, I recall vividly that question was answered. In a now famous paper, South African zoologist Jennifer Jarvis revealed that vertebrates *can* in fact be eusocial.[1] NMRs, which she had been studying in quiet obscurity for fifteen years, were the first known eusocial mammal. Even now, forty years later, only a handful of other vertebrate species can claim eusociality.

Naked mole-rats live astride the equator in East Africa, where the underground temperature is warm and constant throughout the year. Their complex and extensive burrow systems include narrow, shallow foraging tunnels that lead to tubers, some as big as a football. They dig deeper highway tunnels and wider throughways for traveling between foraging tunnels. Then, deepest of all, is the nest chamber, which can be as much as two meters (6.6 feet) underground, where the queen nurses her pups. There is no throne room, but there is a toilet chamber. Tunnels are kept clean. Debris and excavated dirt are transported to shallow tunnels, where workers will open a small chimney to the surface through which they kick excavated soil. The only visible evidence that NMRs live in an area are these above ground "molehills," which if you quietly observe them in the early morning or late afternoon, you may see dirt being flung from them like smoke out of an erupting volcano. If you approach the volcano quietly and carefully, you may even spot the hind legs of the animal doing the flinging. This is the closest naked mole-rats come to appearing above ground, except when an occasional individual will leave its birth colony to start a new one. At these times—earth disposal and dispersal—they are most vulnerable to predators. The queen in all her royal dignity never exposes herself in this way.

In that famous paper describing their eusociality, NMR longevity is mentioned only in passing. Some workers, it was noted, live at least seven years in a captive colony. That is a reasonably long life for an animal the size of a mouse but not really exceptional. The longevity quotient is 1.25. In the late 1980s, when my scientific interests were shifting from behavioral evolution (where I been keeping an eye on naked mole-rats because of their eusociality) to aging and longevity, it occurred to me that NMRs might also be interesting from an aging perspective, too. I contacted Jarvis to ask if she had any updates on the longevity of naked mole-rats, particularly the queens. They live at least into their late teens, she told me, but they still seemed healthy and vigorous at that age. Now I knew their longevity quotient was at least 3.0. The more I learned about them, the more intriguing they became.

Like ants and termites, which also spend much or all of their lives in lightless underground tunnels, it is difficult to study naked mole-rats in their natural habitat. What we know has been largely, although not exclusively, learned from establishing laboratory observation colonies. NMRs, it turns out, take well to captivity. We now know that sterile workers have their sterility enforced by the queen. If you remove her, one of the female workers will transform into a new queen—usually after a bloody fight with aspirant sisters. Removing queens allows new colonies to be quickly and easily created for laboratory research without destroying the old one. In fact, NMRs have become popular exhibits in zoos and museums, where their constant bustling activity in transparent plastic tunnels and chambers has delighted many thousands of people on several continents. A fictional NMR, Rufus, in the cartoon series *Kim Possible* even won a Daytime Emmy Award for Nancy Cartwright, who provided his voice.

Of the species I highlight in this book, the naked mole-rat is the only one that has been seriously studied from an aging perspective. This is due largely to the efforts of Rochelle Buffenstein, a former PhD student of Jennifer Jarvis, who brought some naked mole-rats with her when she moved to the United States in the late 1990s. At about that time, Buffenstein's interests began shifting away from her earlier work on how NMRs manage to get sufficient vitamin D. In most mammals, vitamin D requires sunlight to transform to its active form in the skin. So how animals that live in profound darkness manage to get sufficient vitamin D was the puzzle. But by the time she emigrated, her interests had turned to how they manage to live so long.

In the meantime, the animals that she and Jarvis had captured in Kenya many years before kept on living. Today, the record longevity for a naked mole-rat is thirty-nine years.[2] Thirty-nine years would give them a longevity quotient of 6.7, which is greater than that of humans, nearly ten times the LQ of a laboratory mouse, and even greater than the longest-lived wild bird.

The longevity details are interesting. In Buffenstein's captive colonies, queens and workers differ little in how long they live. The queen, as with queen ants and termites, is the biggest individual in the colony. She remains fertile as long as she is alive. Buffenstein's longest-lived queen so far gave birth to more than a thousand pups in her very fecund lifetime. Perhaps the most striking feature of NMR longevity is that until their later twenties, they appear to show few apparent signs of aging. No gray hair, of course. Wrinkles? Yes, but the young are wrinkled too. Also they show no change in heart function, metabolism, or bone quality. Perhaps most unique, according to current evidence, they do not seem much more likely to die as they grow older, at least into their twenties. This is very different from almost any other species. In animals from insects to elephants, the impact of aging results in the risk of death increasing exponentially with age. Human mortality rate doubles about every eight years after age forty. In mice, it doubles every three months. In NMRs, it may never double at all after they become adults at about six months of age.

In nature, things look a bit different. Stan Braude, now of Washington University in St. Louis, has been studying naked mole-rats in the field since he was an undergraduate in the early 1980s. He reports that the longest-lived queen he has studied in nature lived seventeen years (which would still be a longevity quotient of 3.0) but that workers live only two to three years—a very termite-like difference.[3] Of course, in nature, life in even the best-protected habitat can be dicey. The environmental dangers include snakes, burrow flooding, parasites, pathogens, and squabbling with fellow colony members.

One other thing that Buffenstein reported made biomedical researchers sit up and take notice. NMRs, living more than ten times as long as laboratory mice, didn't appear to get cancer. That observation is stunning and suggests that we may have a lot to learn about cancer prevention from them. However, that observation also requires a short digression into cancer, longevity, their relation to one another, and the ways both are affected by body size.

NAKED MOLE-RATS AND CANCER

Recall that any species with continued cell replication throughout life risks developing cancer. That risk is because as long as cells retain their capability to replicate, they also retain the capability for that replication to lurch out of control. Unbridled cell division is cancer—a price species pay for their ability to continue cell division throughout life to repair or replace damaged parts.

Long life requires the ability to repair external wounds and internal damage, but it also requires cancer resistance. Cancer resistance has evolved by the development of tight, redundant controls on cell division. To use an automobile analogy, this is like having several independent acceleration governors to make certain speed is controlled plus multiple brakes to stop if needed. Cancer biologists call the governors *proto-oncogenes*. They control when cells divide. If they become mutated, they can become oncogenes—cancer genes, in other words—where replicative control is lost. What I am calling *brakes* cancer biologists call *tumor-suppressor genes*. By various means, they suppress out-of-control cell division. When they become mutated, they lose their ability to suppress. Species that live a long time must have more and better governors and emergency brakes than others.

Rochelle Buffenstein, once she got interested in aging, wanted to know what diseases ultimately killed her very long-lived naked mole-rats. When animals died, she would open them up. What she noticed is that she never found tumors, even after examining more than a thousand deceased NMRs. As more and more animals, living in more and more captive colonies, underwent more and more postmortem exams by more and more researchers and veterinarians, ultimately a handful of tumors were found. But they were remarkably rare compared to what was found in old animals of other well-studied species, such as mice, rats, dogs, cats, or even long-lived species like parrots and people. When researchers tried to induce tumors in NMRs with chemical carcinogens that were tried-and-true in mice, they couldn't do it.[4] Here was a striking example of how nature had something important to teach us about reducing a deadly and debilitating human disease.

We don't yet understand how naked mole-rats so successfully avoid cancer, but some clues are beginning to emerge. Husband and wife research team Andrei Seluanov and Vera Gorbunova from the University of Rochester have discovered a novel chemical in NMR skin that may play a role in their cancer resistance. That chemical is a special form of something that we have in human skin as well, called *hyaluronic acid*. A kind of goo that seemingly plays a role in keeping skin pliable, it does a great deal more. In joints and in the membranes surrounding organs, it serves as a lubricant. It is involved in cell division, building new blood vessels and repairing wounds. It is also a component of numerous cosmetics, particularly for treating dry skin. The Seluanov-Gorbunova team has found that the special form of hyaluronic acid that NMRs manufacture can make mouse cells in a dish display some of the same properties possibly linked to cancer resistance that NMR cells display.[5] It is early in these studies, but stay tuned. Time will tell the ultimate importance of this discovery.

There is more, much more, to understanding their exceptional longevity than their extreme cancer resistance. Although we haven't discovered those secrets yet, we have discovered one important secret that it isn't. Remember that byproducts of normal metabolism are oxygen radicals that damage virtually all other biological molecules. Longer-lived species in general have less oxygen radical damage than shorter-lived species at the same age, either because their slower metabolic rate produced fewer radicals or better protection against them. Naked mole-rats buck this trend. At the same age, they have more—rather than less—oxygen radical damage than mice![6] This finding was so unexpected I didn't believe it until it was reported several times using several different ways of measuring that damage. It made more sense when it was discovered that the genome of the NMR lacked one major antioxidant gene. So NMRs have taught us that it is possible to live a long life despite high levels of oxygen radical damage. We don't understand what allows NMRs to be so tolerant to that damage, though, which is another thing that these charismatic rodents may have to teach us.

There is another subterranean-living rodent that may have something every bit as interesting—but different—to teach us about long life and cancer resistance. It is not closely related to the naked mole-rat despite its similar name.

MIDDLE EAST BLIND MOLE-RAT

The Middle East blind mole-rat (*Spalax ehrenbergi*) lives, as you might guess, throughout the Middle East and is profoundly blind, as its name implies. Its rudimentary eyes are covered by skin. It is more closely related to mice than to naked mole-rats. I've mentioned one species name, but *Spalax* is interesting for many reasons, one of which is that it seems to be in the process of splitting into multiple species, so it is technically called a *species complex*. However, it's not quite finished splitting, so I will continue to call it by a single name for simplicity—just *Spalax*.

The ecological and physical similarity of *Spalax* to naked mole-rats is remarkable despite their rather distant evolutionary relationship. *Spalax* also lives underground in complex sealed burrows, eating roots and tubers. It eats so many roots and tubers that, in fact, it is a major pest to potato and sugar beet farmers in the Middle East. Like the naked mole-rat, it digs these tunnels not with powerful forelimbs like many other burrowing mammals but with elongated powerful incisors. Very much unlike naked mole-rats, *Spalax* live a solitary life in their burrows, emerging mainly for brief interludes for mating.

If eusociality is a key to long life, then we would expect *Spalax* to have a rather normal rodent lifespan. If living an entirely subterranean life is the key, then *Spalax* should be exceptionally long-lived. In a major encyclopedia of mammals published in 1999, 4.5 years (a longevity quotient of 0.6) was its maximum reported lifetime under optimal laboratory conditions. So if I had been writing in 1999, I would have concluded that eusociality was the key. However, that conclusion would have been wrong.

Because its blindness could paradoxically teach medical researchers something about the necessary molecular machinery required for sight, *Spalax* became a more widely used laboratory animal in the 2000s. Improved husbandry and greater numbers—thousands of animals have now been followed from life to death in Middle Eastern laboratories—reveal that those early reports were exceptionally misleading. We now know that *Spalax* can live at least twenty-one years in the lab. This is more like an African mole-rat LQ of 2.9. At last report, animals at this age were still alive, so maybe they can live even longer. Even more interesting, like naked mole-rats, they show virtually no signs of aging across their long life, and they turn out to be even more cancer-resistant than NMRs.[7] In fact, to this date, there have been no—none, zero—reports of naturally occurring cancer among the thousands of *Spalax* kept in Middle Eastern research facilities. It is next to impossible to give them cancer, even with powerful chemical carcinogens. The cancer resistance of *Spalax* and NMRs seems to have some differences, too, based on differences of their cells' behavior in a dish.[8] Unlike NMR cells, which resist killing, *Spalax* cells die at the slightest sign of injury. That, combined with an ability to quickly replace dying cells, may be important in explaining their cancer resistance. So there may be different lessons to be learned from these two species.

What is it about subterranean life, then, that leads to exceptional longevity? This is clearly a protected niche, offering escape from some external dangers. But what might cause the protection from internal dangers like cancer? An intriguing idea is that it may have to do with the air they breathe.

IT'S AS EASY AS BREATHING

If you could be magically shrunk to a size that you could walk through and explore the complex burrow system of these mole-rats, your exploration would not last long because the air deep in those tunnels would quickly kill you. It has too little oxygen and too much toxic carbon dioxide and ammonia.

Normal air contains 21 percent oxygen, 0.04 percent carbon dioxide (0.03 percent before the world industrialized), and virtually no ammonia. We are designed to thrive breathing that type of air. However, when you inspire air, you extract some of the oxygen and exchange it for the carbon dioxide, a potentially toxic by-product of metabolism that you need to remove. Expired air is around 16 percent oxygen and 5 percent carbon dioxide. I hope you noticed that expired air has about one-quarter less oxygen but 125 times as much carbon dioxide as inspired air. We normally don't think of carbon dioxide as toxic, but high levels in the body make our fluids more acidic, which *is* toxic. Humans do the opposite of thrive when rebreathing that air with too little oxygen and toxic levels of carbon dioxide. Ammonia I'll ignore except to point out that it is highly toxic at pretty low concentrations. Ammonia toxicity symptoms include coma and death, symptoms that should definitely be avoided. Suffice it to say that numerous animals urinating deep underground in the same latrine would make breathing near the latrine as toxic as it would be unpleasant. Large bat colonies roosting deep in caves also produce toxic levels of ammonia.

We know quite a bit about our tolerance for low oxygen and high carbon dioxide thanks largely to research on how to keep submarine crews alive and alert. Early submarines could stay submerged for only a matter of hours without surfacing to replenish fresh air. Fortunately, we now have the technology to manage submarine atmosphere with oxygen producers and carbon dioxide scrubbers. Research on sailors aboard nuclear submarines, which can stay submerged for months, has found that if oxygen drops below 19 percent or carbon dioxide rises above 0.6 percent, the crew begins to suffer from drowsiness, headaches, fuzzy thinking, and other things that you don't want people in charge of nuclear weapons to suffer.[9] Again, notice that 0.6 percent carbon dioxide may sound like a little, but it is fifteen times the amount in normal air. Astronauts have similar problems. Scott Kelly, the American astronaut who spent a year aboard the International Space Station (ISS), complained in his book that the acceptable standard for carbon dioxide aboard the ISS was three times higher than that aboard nuclear submarines. These higher levels gave him headaches, congestion, burning eyes, and fuzzy thinking.

Animals living in sealed underground burrow systems do not have machines to generate fresh oxygen or carbon dioxide scrubbers. They are stuck with rebreathing the same air, freshened only by whatever small amount diffuses through the soil or ventilates the burrow through molehills briefly opened to dispose of excavated dirt. In foraging tunnels near the surface, the air might be reasonably fresh, but deep down where nests and latrines are, there will be less oxygen and more carbon dioxide. Burrows with many animals breathing in them will be worse. Mole-rat burrows have been recorded to have as little as 6 percent oxygen and as much as 10 percent carbon dioxide. Either of these concentrations would render a human unconscious within minutes and dead not long after. So animals that live in sealed underground burrow systems, like mole-rats, have had to evolve a tolerance for low oxygen and high carbon dioxide.

Naked mole-rats, likely because they live in the largest underground groups, are among the most tolerant mammals known for their ability to manage life with little oxygen and lots of carbon dioxide. Experiments have shown them to display no distress in 5 percent oxygen, which kills mice in less than fifteen minutes. They have survived for hours at 3 percent oxygen and 80 percent carbon dioxide—yes, 80 percent carbon dioxide, two thousand times the concentration in open air. Contributing to their tolerance for low oxygen, by the way, is their ability to lower their already low metabolic rate even farther. It turns out other African mole-rats tolerate low oxygen exceptionally well—not quite as well as the NMRs, perhaps, but very well compared to almost any other mammal.[10] After all, they all do live in sealed underground burrow systems. I suspect they also all have exceptional tolerance to high carbon dioxide, but that not yet been investigated in many species.

What about *Spalax*? They are solitary and so have only one or one plus newborns breathing in their sealed underground chambers. Their burrows are shallower than NMR burrows, but despite that difference, air in their burrows has been measured having as little as 7 percent oxygen and as much as 6 percent carbon dioxide—almost NMR-like. They clearly can tolerate very low oxygen and high carbon dioxide, too.[11]

Could there be a link between the tolerance for low oxygen, high carbon dioxide, cancer resistance, and longevity? I think there might be. The strange atmosphere in underground burrows stresses your cells. There are specialized suites of genes we all have that are turned on when oxygen is low or carbon dioxide high. At the level of the cell, oxygen can be depleted, and carbon dioxide can be elevated to minor extents all the time, such as during exercise or when traveling in the mountains. Animals that live in sealed burrows, though, experience this to a dramatically greater extent and do so for much of their lives. They need to evolve especially effective defenses against or tolerance for oxygen and carbon dioxide stress. Those defenses may also protect against more normal internal dangers such as damage to DNA and other cellular components. Learning the details of how those defenses work, particularly to the extent that they are more effective than the defenses humans already have as they appear to be, could possibly be turned into health benefits for humans in time.

What about other subterranean mammals, like pocket gophers or moles, that are more familiar to us in North America? Do they live exceptionally long lives and have exceptional cancer resistance? The answer is no. Despite their reputations for living underground, compared to mole-rats they do come above ground routinely. Moles, in fact, forage in tunnels that are so close to the surface that you can often see them as bulges in the dirt. So they do not live with the same protection against external dangers as do the mole-rats. Nor do they spend their lives breathing air that would be dangerously toxic to humans.

There is another ecological niche that we may want to consider for its relative safety, though not its lack of oxygen. What about animals that spend their lives in caves?

THE HUMAN FISH

Common names of animals are my secret source of delight. So far, we have encountered the flying lemur (which neither flies nor is a lemur) and various mole-rats (which are neither moles nor rats). Now let me introduce you to the human fish (*Proteus anguinus*), which—surprise—is neither human nor a fish. It is a small, white, blind salamander, also called the *olm*.

Salamanders may be our most seldom seen vertebrates. Resembling slimy-skinned lizards, they tend to be active at night, hiding under rocks or inside logs during the day. Except for species that warn of their toxic skin with bright colors, they tend to be drab, blending into their background. The smallest species is no bigger than the distance between your finger joints, including its tail. The largest species, the Chinese giant salamander, can grow to the size of a small person. While opinions vary on whether the naked mole-rat is cute or hideously ugly, opinions on large salamanders are virtually unanimous. Only a salamander mother could disagree.

Salamanders evolved in central Asia during the time of the dinosaurs, and most species still live in northern latitudes. As ectothermic animals living in cool climates and therefore having low metabolic rates, salamanders might be expected to be long-lived as a group—and they are. Like almost all other groups of animals (except bats), larger species tend to live longer than smaller species, but even a small standard-issue species such as the deliciously named northern slimy salamander (*Plethodon glutinosus*) of the eastern United States, which is small enough to rest comfortably on your palm, can live up to twenty years as a pet. That is a longevity quotient of 5.3, similar to that of the longest-lived wild birds.

The olm is a very special salamander—a cave salamander—meaning that instead of visiting caves only for roosting or hibernation like bats, it spends its every moment of its life inside caves (figure 8.2). Like sealed underground burrows, caves provide a constant, safe environment, particularly if you are at the top of the food chain, as olms happen to be. There are few species that spend their lives entirely in caves. In Europe until recently, the olm was the only known cave vertebrate cave of any sort. Like sealed underground burrows, cave life is one of profound darkness. Nature abhors wasted energy, so living in total darkness obviates any reason for evolution to waste energy on producing skin pigment or eyes—hence the olm's white skin and tiny, functionless eyes. Maybe that explains its puny limbs as well.

The olm lives in the underground ponds and streams inside the extensive limestone cave systems that honeycomb the southern Balkans. Temperature is a constant 10°C (50°F). At this temperature, everything about the olm's life is slow. Eggs take about five months to hatch into twelve-millimeter (half-inch) tadpoles, and olms do not reach adult size and reproductive adulthood for a very human-like sixteen years. Females lay clutches

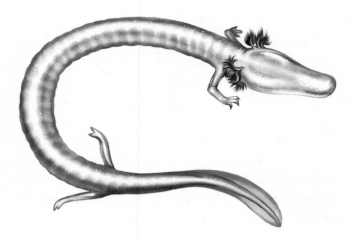

Figure 8.2

The olm or human fish. With its absurdly slow life style, they mate every dozen years or so. One individual was found in the exact same spot it was originally captured after seven years. Adults are about twenty centimeters (eight inches) long, including the tail.

Source: Courtesy of Shutterstock.

of around thirty-five eggs and, being in no hurry, wait a dozen years between clutches. They can also survive at least a year—maybe longer— without eating.

The olm's exceptional longevity has been known for many years. Zoos report them to live up to seventy years. However, a more recent formal mathematical projection from analyses of their survival pattern up to age sixty in nature estimated that their maximum longevity should work out to be around 102 years.[12] Whether you want to believe seventy years or 102 years for their real maximum, given their small size (about the length that your hand can span), those are eye-popping longevity quotients of 14.4 and twenty-one, respectively—bigger than any we have seen before. If you're checking longevity boxes, we have an ectothermic animal (check) living at cool temperatures (check) in a stable, protected environment (check). The only box we haven't checked yet is maybe resistance to low oxygen. Caves typically have multiple openings to the outside world, so being reasonably well-ventilated, the air inside caves does not lack for oxygen. However, the olm lives in the water inside perpetually dark caves. Green oxygen-producing algae or plants cannot survive in the dark. So cave water must get its oxygen sporadically, mainly from rain water seeping in from above. During droughts, cave water may have very little oxygen, indeed. Surprise of surprises, then: olms are exceptionally resistant to low oxygen, too.[13] Check.

So life that takes place in the dark, is oxygen-poor, and occurs underground may have something to teach us about how to live long, presumably cancer-free lives. We will have to wait and see about that. At least researchers are investigating some of these species. What about animals that don't have the advantage of spending their lives in burrows, in caves, with the ability to fly, or with a protective shell around them? Do they ever evolve the ability to live long lives? Let's see.

9 THE BEHEMOTHS (ELEPHANTS)

If size and longevity go together like strawberries and cream, then the largest animals ought to be of interest in any discussion of animal longevity. The largest animals to ever roam the earth are no longer with us, of course. For all the advantages large size may seem to confer, it didn't help the appropriately named titanosaurs, who perished with the rest of the nonavian dinosaurs when that massive asteroid slammed into the earth 66 million years ago. Exactly which of the plant-eating titanosaurs was the biggest is a topic guaranteed to start a food fight at a paleontology convention. Whichever species it was, it weighed somewhere in the vicinity of fifty to a hundred tons and was thirty-ish meters (a hundred feet) in length, with a brain no bigger than a tennis ball. So whichever species it was, it wasn't all that smart and probably didn't live all that long. Remember, the longest-lived dinosaur known so far was the relatively puny twenty-ton *Lapparentosaurus madagascariensis*, which survived only into its forties.

Size didn't help the largest mammal to roam the earth, either. This was probably a rhinoceros-like beast called *Paraceratherium*, which stood almost five meters (sixteen feet) high at the shoulder and weighed fifteen to twenty tons. *Paraceratherium* vanished around 25 million years ago for reasons that aren't clear, although, thank goodness, it was too early to have been done in by humans, like many of the more recent large mammals. We have no idea how long it might have lived, as its bones, like the bones of other mammals, do not have adult growth rings like many dinosaurs, although modern molecular biology is providing us with tools that may someday offer us a way to estimate the age of animals dead for millennia.

The largest terrestrial mammal alive today is the African bush elephant (also called the *savanna elephant*) (*Loxodonta africana*), which can reach four meters (thirteen feet) height at the shoulder and weigh up to seven thousand kilograms (15,500 pounds). Unlike dinosaurs and long-extinct mammals, we do know something about elephants, including how misleading it is to use any single number to describe their size or their lifespans.

The first identifiable elephant ancestor arose during the explosion of mammal diversity that followed the great extinction of 66 million years ago. Initially no bigger than a house cat, evolution favored increasing size in this group until we find the biggest elephant of all time, the Asian straight-tusked elephant (*Palaeoloxodon namadicus*), which was half again as tall as today's species and weighed four to five times as much. The end of the ice ages was also the end of *Palaeoloxodon*, as it was for its more familiar relatives, the wooly mammoths of the far north and the more temperate mastodons, a species most famous for being the primary seed disperser of the ancestor of what became today's Halloween pumpkins. The closest living relatives of the three elephant species alive today are the seagoing, slow-moving, vegetation-eating dugongs and manatees.

If you are surprised that there are three—rather than the traditional two—species of elephants alive today, you are not alone. Asian elephants (*Elephas maximus*) originated in Africa, splitting off from today's African elephant species about 5 million years ago. In historical times, they have been largely confined to South and Southeast Asia, as the name implies. It is the small-eared, easily trainable species used today for public entertainment in circuses and various kinds of labor. Then there is the larger, large-eared African bush or savanna elephant (*Loxodonta africana*) found in, you guessed it, the sub-Saharan African savanna. Bush elephants, being less docile and more difficult to train, do not work for peanuts. Finally, there is the more recent, possibly new to most readers, African forest elephant (*Loxodonta cyclotis*), which was officially recognized as a separate African species in 2021, when genetic evidence dictated it should be. This is the smallest of the three species, found only in West and Central African rainforests. As forest elephants are poorly known compared with the other two species, I won't intrude further on their privacy here.

There are no more iconic animals than elephants. Schoolchildren world-wide recognize the huge animal with its distinctive tusks and trunk, whether or not they have ever seen one in person. It has been this way for millennia. Elephants are depicted in Stone Age rock art. They play major roles in Hindu and Buddhist shrines and temples and figure in many indigenous religions. They have been kept captive and trained to work for humans at least since 3000 BC. They have been used to pull chariots and as weapons of war since ancient Greek times. The Indian emperor Chandragupta is said to have owned nine thousand war elephants. They devastated Alexander the Great's army during his Indian campaign of 326 BC. Asian royalty has ridden on elephant backs to hunt and flaunt their superiority for as long as royalty has hunted and flaunted its superiority. Today, they still work in the timber industry in Southeast Asia. Wherever wild elephants and farmers live together, there is friction over which species has most right to the crops. Humans call elephant behavior crop raiding. Elephants no doubt think of it as drive-through dining. Wherever they are known (other than by irate farmers), they represent strength, wisdom, loyalty, power, leadership, and even, of course, longevity.

They are among the most intelligent of mammals, no matter how you define intelligence. They are fabled for their long memories, as we all know, but more than that, they can learn to use a variety of tools. That trunk, by the way, is quite the tool itself. A fusion and extreme elongation of the elephant's upper lip and nose with no bones or joints and as many as sixty thousand individual muscles, it combines strength, flexibility, and remarkable dexterity. Like our nose, it is used for breathing and smelling the roses. When trudging through deep water, it is a snorkel. Unlike our nose and more like our hands, it is also used for grasping, lifting, eating, drinking, bathing, and self-examination. It can lift as much as four hundred kilograms (880 pounds) and delicately shell a peanut. I once had my wallet plucked from a back pocket by a mischievous juvenile African elephant. A fun fact you may want to file away for future use is that African elephants have two small muscular projections at the tip of the trunk which they use for fine grasping as we might use a mittened hand and thumb. Asian elephants have only one projection, which they wrap around objects they wish to grasp—a thumbless mitten. For tools other than their trunk, they have been seen modifying branches to use as fly swatters and back scratchers, to plug holes they have dug, and to reach objects too high for their extended trunk.

Tusks also can be thought of as tools. They are used for digging, scraping, and clearing brush and in combat. Like human "handedness," individual elephants favor using one tusk over the other, which means one will be more worn and shorter than the other. Both sexes have tusks in African elephants, although males' tusks are typically thicker. In Asian elephants, with rare exceptions, females lack them.

Tusks are teeth, no different from yours and mine except in size and shape. If your teeth were as big as elephant tusks, you might be poached for your ivory, too. Tusks are modified incisors, and like rodent incisors, they grow throughout life. In the past, tusks might reach three meters (ten feet) in length and weigh close to a hundred kilograms (220 pounds), but selective hunting of exceptionally large elephants for their exceptionally large tusks over the generations has genetically reduced tusk and probably elephant size. Ivory—whether from an elephant, a walrus, a hippopotamus, a sperm whale, or a human—has a lovely creamy white color, is durable, and can be carved without splintering. Since ancient times, it has been carved into a variety of decorative objects. In modern times, its nondecorative uses have been as billiard balls, piano keys, dominoes, or, in splendid irony, so-called false teeth. Human lust for ivory has led to the collapse of elephant populations worldwide. This was apparent from the early nineteenth century, when a New York billiard ball supplier, realizing that elephant ivory could disappear, offered a $10,000 prize (about $200,000 today) to anyone who could come up with a substitute for elephant ivory. John Wesley Hyatt did so in 1869, although whether he got the prize money is unknown. Today, anything that can be made from ivory can be made from synthetic materials, and international trade in ivory has thankfully been banned.

Their intelligence allows elephants keep a mental map of large areas. They remember the location of water holes and food patches and the best route for getting to them. They also remember and avoid sources of danger. They recognize individual family members by sight or scent and can remember where they last encountered those family members. They even remember the scent of dead relatives who have been gone for years. They can figure out how to cooperate with one another to perform tasks that single individuals could not. They can be taught to recognize numerous individual objects and have some notion of numbers. In circuses, elephants learn dozens of tricks, including things they would never do in the wild, like stand on their heads. They recognize themselves in mirrors, showing self-recognition, in other words. This is considered by comparative psychologists a sign of particularly sophisticated intelligence that only humans, other great apes, dolphins, Eurasian magpies, cleaner wrasses (fish), and octopuses possess. Okay, the octopus results were equivocal. Most impressive to me is that elephants can distinguish from voices alone (perhaps recognizing human languages) people from tribes that hunt elephants from those that do not.[1] Maybe none of this should be surprising for an animal with a brain three times the size of our own.

Elephants' great size has benefits. One obvious one is that once they reach adult size, like giant tortoises, they are virtually immune from predators other than humans, even though they live in the midst of plenty of large predators, like lions, leopards, hyenas, and tigers. They also get their way at critical resources such as water holes, where lions, leopards, and even hippos give them plenty of room. They understand the damage their size can do and are acutely aware of where they place their feet and how much weight they put on them. Elephant calves standing amid a tightly packed protective circle of adults are in no danger from their protectors. On the other hand, whatever animal is threatening them very definitely is. In Aberdare National Park in Kenya, years ago, researchers spotted a rhinoceros attacking an elephant calf and also spotted it being summarily stomped to death by the calf's adult relatives. When I worked as an animal trainer in the movies, my foot once served as a stunt double for the foot of a well-known actress who played the Bionic Woman. Being bionic, in the script she was supposed to not notice that an elephant was standing on her foot. Unwilling to risk injury to his star, the director gave my foot the part. That foot—the right one—helped finance my graduate education. During the shoot, incidentally, the elephant indeed touched his foot on mine as lightly as any cat.

One cost of their great size is that it limits the way elephants can move. They have limited mobility in their knees and elbows. They can't trot, gallop, jump, or rotate their front legs. Any of these movements would risk damaging their knee or elbow joints or breaking their legs, which are supported by the thickest leg bones of any animal. The stresses of even a small jump or twist could be too much. Their run looks more like a racing walk—a really fast racing walk—so don't imagine that you can outrun an elephant. Gravity can be the enemy of great size. A major cause of death among elephants in the timber industry is falls. It is only in Dr. Seuss stories that elephants think of climbing trees. Living in the wild on exceptionally low-calorie, high-fiber food (grasses, branches, and bark), they spend up to 70 percent of their waking hours eating and need to consume 100 to 180 kilograms (220 to 400 pounds) of their low-quality forage daily to provide enough energy to support their great bulk.

Although some details vary among the three species, elephants in nature live in tightly bonded social groups of adult female relatives along with their calves and juvenile offspring. As females mature, they stay in the family group, but when males near adulthood, they gradually spend less and less time with their family, finally departing for a solitary life or a loose social life in bachelor herds. They may join female groups from time to time in the hope of finding a fertile female ready to conceive. That seldom happens. In Amboseli Park, Kenya, where elephants have been studied continuously for nearly fifty years, researchers calculate that each female is fertile only three to six days every three to nine years. Aspiring fathers need to be alert and ready.

Adult males know the other males in an area, and dominance hierarchies exist based on size, for the most part. I say "for the most part" advisedly because male elephants display a mysterious feature not seen in other mammals. That unique feature is called *musth*. Musth is a considerable danger to trainers and keepers of captive elephants. It can also pose a danger to other animals unfortunate enough to be in the vicinity. Musth is sometimes thought of as a sexual frenzy, although I think of it more as sexual crankiness. In George Orwell's famous short essay on the evils of colonialism called "Shooting an Elephant," he, as a young colonial policeman, is asked by frightened Burmese villagers to shoot a usually tame elephant who has gone a bit crazy during musth. Having broken its chain—captive elephants in musth are often chained up until it's over—it had destroyed a home and several fruit stands and killed a cow and an unfortunate man who was in the wrong place at the wrong time.

It is difficult to predict when musth will occur, especially in younger males. Female presence is helpful but not required. You can spot a male in musth by the heavy secretions dripping down his cheeks, by odorous urine dripping from his penis, and from his aggressive attitude toward pretty much anything. When in musth, a male's testosterone level shoots through the roof. He temporarily races to the top of the dominance hierarchy, no matter his size. I think of this as "watch out for that dude, he's crazy" syndrome. Being in musth is hard work, though, and males visibly lose weight during it. In nature, males in musth guard receptive females even though it costs them foraging time because, after all, what female could resist a cranky

---

Box 9.1

What Happens When You Give an Elephant LSD?

Possibly the craziest so-called science experiment ever had to do with elephants and musth,[2] and although it doesn't directly pertain to elephant longevity, I can't resist mentioning it here. It was published in one of the world's preeminent scientific journals in 1962, and all I can say about that is that it shows how far science has come since 1962.

The researchers were ostensibly trying to understand musth as a service to protect captive elephant keepers. How might you do that? Well, in the 1960s, you might think, "Let's give this male elephant the psychedelic drug LSD and see if it makes him behave like he's in musth." Groovy, man.

The procedural problem was how much LSD to give an elephant if you wanted to see if it made him crazy? Would you use a dose that in humans led to "vivid visual hallucinations and grossly disorganized psychotic thought and behavior" or the higher dose required to produce "a transient rage reaction" in a cat? Okay, let's go with the cat dose. The elephant weighs about three thousand times as much as a cat, so let's give it three thousand times the amount a cat gets.

Maybe they should have remembered how metabolism and pretty much everything else slows with body size, but they didn't.

Unsurprisingly, things did not go well. Soon after receiving the drug, the poor elephant named Tusko began to sway, his legs buckled, he defecated, collapsed into seizures, and within two hours was dead. No sign of musth. Although I personally enjoyed them, the 1960s had their problems.

---

male with odorous urine dripping from his penis and fluid running down his cheeks? Most elephant calves, however, are sired by males in musth.

As might be expected for animals as large as elephants, it takes a while to grow into adulthood. They spend nearly twenty-two months developing in the womb, the longest time of any terrestrial mammal, and weigh around a hundred kilograms (220 pounds) at birth. They suckle for another three years or so, and what happens after that depends a lot on the amount and quality of food they find and when their mother has her next calf. In most animals, the rate of adolescent growth and timing of reaching of sexual maturity is exquisitely sensitive to energy balance—the amount of food eaten compared to the amount of work required to get that food, as well as general health. This is most obviously true in humans, where energy balance and diet quality can have an enormous effect on age at maturity. The age of first menstruation, called *menarche*, is particularly easy to document. Age at menarche varies dramatically across cultures, ranging from sixteen years in places where life is energetically difficult (such as Senegal and Bangladesh) to around twelve years in western Europe and the United States).[3] The exceptions help prove the rule. Adolescent girls involved in high-intensity physical training (such as gymnasts and swimmers) delay puberty relative to their less active friends. This may be because a critical percentage of body fat may be required for sexual maturation. The same trend of earlier sexual maturity in recent times is true for adolescent boys, although boys overall mature later than girls.

To understand adolescent (and adult) growth, reproduction, and longevity in elephants, we need to consider Asian elephants and African elephants separately. Neither species thrives in zoos. They don't reproduce as reliably or live as long as elephants in the wild. The two species grow, mature, and even age a bit differently. Most important, although we know a lot about both species, we have learned what we know about each from very different circumstances.

ASIAN ELEPHANTS

Most of what we know about the long lives of Asian elephants comes from semicaptive animals used in logging. Because they are comparatively docile and highly trainable, elephants have been essential employees of timber companies in Southeast Asia since the nineteenth century. There were many more elephants then. Their numbers have plummeted in recent times due to habitat destruction, conflicts with people, and their valuable tusks. The timber camps of Myanmar (known as Burma until 1989) use them to selectively harvest valuable teak trees (*Tectona grandis*) to make elegant furniture for rich Westerners.[4] They do much less damage to the forest than mechanized harvesting would do. They have been called nature's greatest and most generous gift to Burma, although their use in logging has declined in Burma and elsewhere in recent years.

Semicaptive means their owners do not feed or selectively breed them. After a day's work, they are released unsupervised into the forest to forage on their own. At that time, they can mate, fight, or socialize with other timber elephants or with wild elephants. Unlike what you may hear from well-meaning, low-knowledge animal rights activists, timber elephants, like circus elephants, are generally treated well. Whenever good animal health confers economic benefits to their owners, animals tend to be treated well.

In the Burmese timber industry, elephants do arduous work. They haul downed trees through difficult muddy and hilly terrain. They push logs with their heads and drag them along paths to the river, where the logs float to market. They even break up logjams in the river if necessary. Their workloads are carefully managed. They are not given a full load until they are adults. The type and amount of work they do and their rest periods are based on age, size, and physical condition. In fact, they do not work at all in the hottest, driest times of the year. They get reduced workloads as they begin to age and are retired at age fifty-five, living out the rest of their lives in well-earned leisure. All in all, this may be a better deal than many humans living in Burma get.

Most important for understanding the history of their lives and their longevity, each of the thousands of animals trained for timber work in Myanmar has by law its own registration number and name. Meticulous records are kept on their date and place of birth as well as when and how they died. If they were born in the wild, their estimated age as well as the place and method of capture is recorded. Estimating the age of young elephants is highly accurate if done by experienced workers. Studbooks containing records of thousands of animals are kept by veterinarians who must make annual reports to the government, the official owner of all timber elephants.

So what have we learned from these thousands of records of semicaptive Asian timber elephants?

The sexes have very different life trajectories, for one thing. Adult males or bulls (as they are called) are about 10 percent taller and 30 percent heavier than cows (as female elephants are called). But the size difference increases with age. The female life trajectory is eerily reminiscent of modern humans, despite their much larger size. Females reach physical sexual maturity, the elephant equivalent of menarche or first menstruation, at ten to twelve years, although (like adolescent girls) they are not very fertile at that age. Burmese timber elephants are most fertile by about age twenty. Like humans, most females stop reproducing in their thirties, but some continue on. Elephants can continue on even longer than people because they have no menopause. A few still have calves in their sixties.

I don't want to paint too rosy a picture of elephant life in the timber camps. There are stresses, particularly among those elephants who were born in the wild and captured later. Timber companies typically capture elephants in adolescence at an average age of eleven years, but some are as young as five, and others as old as twenty. The effects of capture trauma are apparently lifelong as these captured animals do not reproduce as well or live as long as animals born in captivity.

Infant mortality numbers in the timber camps may appear high at first glance. About 10 percent of calves die before their first birthday, and between a quarter and a third die by age five. To put such infant mortality rates in context, though, this is about the same rate as human infants in Europe during the eighteenth and nineteenth centuries. Also, this is similar to the infant mortality rate in wild African elephants and, consider this, about one-third the rate found in zoos.

For male Asian elephants, the story is a bit different. Male physical development is a bit slower. Males experience their first musth in their early teens but typically don't successfully breed before about age twenty-five. They approach their final height in their twenties but continue to pack on weight throughout their lives, possibly because they continue to grow in length. Size is potentially critical for understanding male lives because size is related to dominance and dominance is related to who gets to mate—at least in African elephants. We assume this is true in Asian elephants as well, but because virtually all of our knowledge about this species is learned from *tamed and trained* elephants living in artificially assembled groups, normal social relations are crazily disrupted.

What do we know about how long Asian elephants live? As with other long-lived popular zoo animals, there are some colorful if dubious biographies associated with extreme elephant longevity. Probably the most famous of these is Lin Wang, a male who reputedly served in the Japanese army, hauling artillery during World War II. The story goes that he was captured during a Chinese victory over the Japanese in Burma. After a heroic trek from Burma to China during which half of the other elephants died, he worked at a variety of jobs, including helping build war monuments and raising money for postwar famine relief. During the Chinese communist takeover in the late 1940s, he was evacuated to an army base on Taiwan. The Taiwanese army eventually donated him to the Taipei Zoo, where he and his colorful biography became their most popular exhibit. Every year from 1983 to his death in 2003 at the reputed age of eighty-five years, the zoo threw Yin Wang a birthday party attended by thousands of paying customers and local politicians.

Okay, I sound skeptical. Why? For one thing, elephants are not typically issued identification papers outside of modern zoos or timber camps. Remember Little Princess, the 157-year-old zoo elephant who miraculously transformed from an African to an Asian elephant sometime during her life? Assuming that the Lin Wang who died in the Taipei Zoo in 2003 was the same elephant who performed all these deeds under all these chaotic circumstances traveling to all these places across all this time requires more faith than I can muster. Also, when we first heard of Yin Wang, he would have already been twenty-six years old. There are no clues about where he came from before the war and how anyone knew his age. Finally, he was a he—that is, a male. One thing we also know about elephants is that, like people, females live longer than males. In almost every debunked case of human longevity exaggeration—and these are legion, as we shall see—the people claiming extreme age were men. It appears the same is true of elephants.

Seriously, here is what we actually do know about the longevity of semicaptive Asian elephants. They live substantially longer than elephants in zoos.[5] As I note earlier, elephants do not thrive in zoos. Also, cows die at lower rates than bulls throughout life and not just in old age. Again, this mirrors closely the human pattern. About one in ten reproductive females lives into her seventies, and the oldest known female with an unassailable birth record lived just a few months short of eighty years. Once again, this is probably close to as long as the oldest humans lived during our hunter-gatherer history. Because of the Asian elephant's size, though, its longevity quotient is only about 1.7. Elephants live a bit longer than expected for their size but nothing like a long-lived bat or a mole-rat. On the other hand, in terms of absolute number of years lived, Asian elephants are second only to humans among terrestrial mammals. Given their great size and this substantial longevity, they must have outstanding defenses against cancer—much better defenses than humans, in fact. We will take that up shortly. In the meantime, how does what we know about Asian elephants compare with what we know about African elephants?

AFRICAN ELEPHANTS

In historical times, African elephants roamed throughout the continent from the Mediterranean to the Cape of Good Hope. At one time, there may have been more than 20 million of them, but like Asian elephants, their numbers have plummeted because of habitat loss, conflicts with people, and human lust for ivory.

The lives of African elephants have been studied extensively under fully natural conditions. The longest-running and most thorough study began in 1972 and continues today in Kenya's Amboseli National Park.[6] The size difference between the sexes is greater in African than Asian elephants, and the magnitude of that difference increases with age. Unlike most mammals, both sexes continue to grow throughout life, but males grow more and faster. A forty-year-old male can be as much as 30 percent taller and twice as heavy as his twenty-year-old self.

Females, like Asian elephants, have a reproductive life not all that different from humans, despite their much greater size. They typically have their first calf in their early to mid-teens and are most fertile from about age twenty to forty, after which their fertility begins dropping progressively. Also like Asian elephants, a handful of females still are giving birth in their sixties. Infant mortality in the wild is comparable that that in the Burmese timber camps and to humans in eighteenth- and nineteenth-century Europe.

Male African elephants, on the other hand, have a reproductive pattern that is different from virtually any other mammal. In most mammals, young adult males are at the peak of strength and ability to hoard access to females, so they father most of the kids—the young stud phenomenon. However, African elephant bulls continue to grow and increase dominance over other males with age. They also have longer periods of musth as they get older, further increasing their dominance. Females appear to prefer older males, too. So males reach their reproductive peak in their late forties to early fifties. Maybe we should call this the old stud phenomenon. African males don't even typically experience their first musth until about age thirty. Old studs pay a price, though. Remember musth is physically draining. Males visibly lose weight during musth, and because older elephants spend more time in musth, it hits them particularly hard. Most males do not live long enough to reach the age of peak fertility.

After fifty years of study, we have complete enough information on African elephants to calculate their life expectancy, something that we don't have for many wild species. The life expectancy *at birth* of males dying of natural (that is, not human-induced) causes is 37.4 years. About 30 percent of males survive to their fifties, and so far none appear to live longer than their mid-sixties. As in humans, extreme longevity seems to be a female game.

Now that I've introduced *life expectancy* and plan shortly to compare elephant life expectancy to that of human hunter-gatherers, I should explain a little about what it is and what it means—and doesn't mean. In essence, life expectancy is the average age at which individuals die. If not specified, it usually means life expectancy *at birth*, roughly meaning the average age of death of all individuals, even newborns. But life expectancy can also be calculated from other ages. For instance, life expectancy at age fifty means the average age of death of all individuals who lived to at least age fifty.

Like any average, life expectancy is most meaningful—that is, representative of something common—when ages follow a bell-shaped curve. In that case, the average number and the most common number are the same. It is less representative of adult deaths when small numbers are common—that is, when there are a lot of infant deaths. This skews the average toward younger ages, so life expectancy no longer says much about age-related deaths. This is important because until the latter half of the twentieth century, human deaths were most common among the very young. That is still true today in nontechnological parts of the world, including among modern hunter-gatherers. All this is to say that if I report that male elephant life expectancy (at birth) is 37.4 years and life expectancy (at birth) among male Hadza hunter-gatherers of east Africa is 30.8 years, it doesn't necessarily mean that elephant adults live longer than Hadza adults. In fact, that is not true. Life expectancy for those that reach at least age fifteen, for instance, is higher among the Hadza, and the oldest Hadza live longer than the oldest elephants. What it does mean is that infant mortality is higher among the Hadza than among elephants. This is an important distinction when discussing the history of human longevity. However, for now, note that male elephants and male humans, both in a state of nature, do not differ greatly in the *average* age at which they die.

How long do female African elephants live, then? For those dying of natural causes in a natural population, life expectancy at birth so far is a very respectable 46.7 years, and the oldest was estimated from tooth wear to have lived seventy-four years (figure 9.1). According to Phyllis Lee, who has been involved in the Amboseli elephant project since the early 1980s, this seems a reasonable number to approximate the limit of African elephant life in the wild. So African elephant longevity at its most extreme (seventy-four years) seems within a stone's throw of Asian elephants (eighty years), making them the third-longest-lived terrestrial mammal. That, by the way, is a longevity quotient of only 1.6, slightly less than the 1.7 of Asian elephants because

Figure 9.1
Barbara, one of the oldest-known African elephants. Part of the Amboseli elephant project, she was estimated to be sixty years old when this photo was taken. She died in 2020 at the estimated age of seventy-two years.
Source: Photo courtesy of Phyllis C. Lee, Amboseli Trust for Elephants.

they are bigger. Elephants live longer than an average mammal of their size (assuming there were any others their size), but nothing to write your local newsroom about.

I can't leave the topic of elephant longevity without at least a brief mention of their teeth—yes, their teeth. Tusks aren't the only unusual teeth in an elephant's head. They also need to chew their food. We humans have two sets of teeth. Our "baby teeth" are replaced by adult teeth that emerge from above, forcing out the baby teeth. Elephants have multiple sets of teeth too. They chew mainly with one upper and one lower tooth on each side. As their food—grass, branches, leaves, bark—is particularly abrasive, these teeth wear down. However, elephant teeth are replaced from behind rather than from above. New chewing teeth emerge in the back of the jaw and gradually push out the teeth in front of them like they were on a conveyor belt. Here is the link to longevity. Elephants have only six sets of chewing teeth. Four of which are typically lost by the time they become adults. By their early forties, their fifth set usually goes, so for the rest of their life, they must rely on that sixth and final set of teeth for chewing their abrasive food. When those are worn way, eating gradually becomes more difficult. Soup and dentures are not available. Teeth, or rather tooth wear, may be what limits the lifespan of elephants.

ELEPHANTS AND CANCER

Elephants weigh fifty to a hundred times as much as humans do. They therefore have roughly fifty to a hundred times as many cells as we do to potentially turn cancerous. To a first approximation, they live about as long as humans, especially preindustrial humans. So do they get lots of cancer? If not, is there something we humans can learn about cancer prevention from them? Possibly. It turns out that elephants have a unique type of cancer prevention.

DNA, as noted earlier, is a language consisting of a very long sequence of four DNA "letters." The sequence of those letters determines whether you are a pea plant, a peacock, or a person and also determines what sort of pea plant, peacock, or person you are. As the price of genome sequencing dropped, we began sequencing the DNA of more and more species, and by now, we have done more than a hundred mammal species, including especially charismatic species that happen to be extinct, like Neanderthal humans, wooly mammoths, and mastodons. Our three-billion-letter genome size is pretty typical for a mammal. The smallest mammalian genome, about a third our size, belongs to a bat; the largest, about three times our size, belongs to a rodent. Yes, that rodent is the famous red vizcacha (*Tympanoctomys barrerae*) of Argentina.

Recall that we have a number of genes called the *tumor suppressors* that are our last line of defense against cancer. The best studied and probably most important tumor-suppressor gene, the one that the elephant has something to teach us about, goes by the underwhelming name TP53—*TP* for tumor protein and *53* describing its size.

More colorfully called "the guardian of the genome," TP53 detects damage to a cell's DNA and orchestrates its response. If the damage is reparable,

TP53 will pause cell division to allow time for the DNA repair machinery to do its job. If the damage is too extensive to be repaired, it flips the suicide switch or hands the cell its cyanide capsule—whichever analogy you prefer—and the cell puts its affairs in order and dies a quick, clean death. Suicidal cells can be replaced by the replication of other, undamaged cells.

The importance of TP53 in cancer prevention can be seen in rare people who have the misfortune to be born with one mutated (that is, inactivated) copy, a condition called Li-Fraumeni syndrome. People with Li-Fraumeni syndrome develop a variety of childhood cancers and have a lifetime risk of cancer that is 73 percent in men and nearly 100 percent in women. Most of us start life with two intact copies of TP53, but they can mutate over time and repeated cell division. More than half of all human cancers have mutated TP53.

It turns out that elephants do not get more cancer than people. If anything, they get less. I qualify that statement because elephants do not survive well in zoos and almost all postmortem veterinary exams on elephants have been done with zoo animals. Cancer is a disease of aging, so when animals do not commonly survive to ripe old ages, they will have less cancer. People in 1900 had less cancer than we do today, not because we live in a more polluted environment today but because they didn't live as long. In any case, elephants clearly do not get *more* cancer than people.

How do they do it? We have one copy of TP53 inherited from each parent. African elephants have twenty copies, the same one that all mammals have plus nineteen backups. Actually, some of the backups have disabling mutations, so there aren't fully nineteen functional ones. The advantage of having multiple backups for genes critical to cell protection seems to be an advantage that nature discovered long ago.[7] If TP53 is the guardian of the genome, their genome is guarded by a small army. Smaller Asian elephants have twelve to seventeen extra TP53 copies. Functionally, this has been worked out by studying elephant cells in a dish and purposely damaging their DNA. It turns out that, compared to human cells, lower levels of DNA damage to elephant cells activates TP53 and flips the cellular suicide switch. Elephant cells are suicide-prone to an exceptionally extent. We know that the extra copies of TP53 are responsible for suicidal elephant cells because researchers inserted those extra TP53 copies into mouse cells, which made them more suicide-prone. This is the type of detailed experimental work that needs to be done on other long-lived species if we are to use their lessons to improve human health. Being satisfied with sequencing genomes is not enough.

Nature is endlessly inventive. We don't have the complete story on elephant cancer resistance yet, but we have a good beginning. Interestingly, blind mole-rats—which are also exceptionally cancer-resistant, due at least in part to easily induced cell death—do something different. We also don't have all the answers there, but they do not have extra copies of TP53, and they do have some changes in their TP53 gene that seem to be responsible for their cancer resistance. Whales—also by virtue of their body size and longevity— have to be even more exquisitely cancer-resistant than elephants. They also do not have extra copies of TP53. Nature apparently has a different trick up its sleeve for whales.

The biochemist Leslie Orgel was fond of saying evolution is smarter than you are. He meant, of course, that because evolution has tinkered with probably trillions of animals over billions of years, its tinkering will have discovered ways of doing things that humans are not likely to work out by deductive logic. Cancer resistance is a problem that nature has been working on for at least 600 million years, and by studying its success stories—like giant tortoises, elephants, whales, and mole-rats—we may discover some novel tricks for boosting our own cancer resistance.

10 BIG BRAINS (NONHUMAN PRIMATES)

Humans are proud of our big brains. We're proud of our opposable thumbs, too, although we share that trait with most other primates and assorted other tree-climbing mammals. The koala, in fact, goes us one better by having two thumbs on each hand. Most primates also go us one better by having an opposable toe, too. So while our opposable digits may not be unique, our brains and the brains of our closest relatives, monkeys and apes, are unique. Ours, for instance, has given us the ability to do amazing things. We have risen from the African savanna to dominate the earth, the air, and the ocean, and we are working on dominating space. It allowed us to invent art, agriculture, and technology. That technology permitted us to transform the earth for better or worse. It now allows us to put the world's libraries in the palm of our hand, transmit information around the globe in the blink of an eye, and transmit it back from the outer solar system in a few hours. We have even developed the capacity to poison the planet and exterminate ourselves and all other creatures on earth, an achievement that maybe shouldn't make us particularly proud. Our brain did not arise *ex nihilo*. We inherited (and then expanded) it from our primate ancestors. Primates were the first evolutionary experiment in big mammalian brains.

All of our closest primate relatives have large brains relative to their body size, the largest of any mammalian group. All are also long-lived for their body size. That is, primates have the highest longevity quotients of any major group of mammals other than bats. This pattern—large brain for size, associating with long life—has led some researchers to speculate (again, based,

I suspect, on our obsession with our own large brain) that a large brain *is responsible for* long life not just in primates but perhaps in other mammals as well.

There is a hint of plausible logic to this idea. Assuming that the brain directs most of the body's responses to various external and internal challenges, maybe bigger brains are better directors, allowing us to make smarter, more finely tuned responses with more precise control. Perhaps they allow us to better anticipate and thus avoid future challenges. However, it is also easy to overestimate what the brain can do with respect to longevity. It is easy to understand that the brain might play a role in timing and success of major life transitions, such as from suckling babe to independent child or from adolescent to reproductive adult. These require coordination of changes throughout the body. In fact, we know that hormones produced by the brain are involved in the timing of these events. However, adult longevity is about prevention and repair of damage at a cellular level as well as about coordination and communication among body parts and good decision making. At least part (possibly the largest part) of a long life is due to cellular properties that can be observed and studied in a dish with no brain in sight. The resistance of a mole-rat or a human skin cell to cancerous transformation does not depend on the brain.

Brain size, as a general rule, is related to body size. A mouse cannot have a brain the size of an elephant no matter how useful it might be. The proportional size of a typical mammalian brain relative to the size of its body decreases in a predictable way as species' size increases. For example, mouse and human brains both weigh about 2 percent of their total body weight. But mice have a smaller than average brain for their size and humans have a much larger one for their size. This is also true of liver and kidney—but not heart—size. Hearts always weigh about half a percent of total body weight whether you are a mouse, a moose, or a sperm whale.

So does having a bigger than average brain relative to its body size mean that a species will live longer than average? The idea of relating relative brain size to other things—primarily intelligence—years ago led to the development of something called the *encephalization quotient* (EQ for short). In fact, I stole the idea of the longevity quotient (LQ) pretty directly from the idea of EQ. So analogously to LQ, EQ is simply brain size of a species relative to average brain size of a mammal of the same body size. It was formulated by paleontologist Harry Jerison in the hope that it would be useful for comparing intelligence across species, living and extinct. Like LQ, an average mammal, by definition, has an EQ equal to 1.0. Greater than that means bigger than average, and less means smaller than average.

Let me be specific that we are considering *species* differences in brain size here, not individual differences within a species. Women have, on average, smaller brains than men because they have, on average, smaller bodies than men. Needless to say, this does not mean that women are less intelligent than men or that men or women with larger brains than others are more intelligent. Albert Einstein had a slightly smaller than average brain, and the Nobel Prize–winning French writer Anatole France had one of the smallest brains ever measured in a full-size human. The nineteenth-century idea that individual differences in human brain size are a measure of intelligence should be dead and buried in the same pseudoscientific grave as astrology, phrenology, and reading entrails to predict the future.

Having said that, might relative brain size indicate *something* about species longevity, and if so, what? Primates as a group have brains that are nearly two-and-a-half times as big as expected. Average primate encephalization quotient is 2.3, to be specific. Also, as a group, they live about twice as long as expected for their size. So far, so good. However, to throw some quick cold water on the generality of this idea, bats, with by far the highest longevity quotient of any mammal group, have roughly average EQs. With my graduate student at the time, Keyt Fischer, I investigated this issue not long after I got interested in aging. We looked at how both absolute brain size and relative brain size related to the longevity of many dozens of mammal species.[1] We also did something that previous cerebro-centric researchers had failed to do. To make sure that we were looking at something unique to the brain, we also examined how other organ sizes—heart, liver, kidney, and spleen—related to longevity. All of these organs, it turns out, correlated with species longevity because all were larger in larger species and larger species live longer on average. The brain had no better (in fact, a slightly worse) correlation with longevity than the other organs—except among primates. *Relative* brain size and *relative* longevity are not linked in mammals generally, but within primates, they very definitely are. For primate longevity, brain size, for some reason, seems to matter.

PRIMATE ORIGINS

Primates arose from small, nocturnal, arboreal ancestors. Their large brains may have been an evolutionary innovation to deal with remembering where patches of favorite foods would appear at certain times of the year, remembering where danger usually lurked, understanding how to navigate in the complex, three-dimensional structure of a tropical forest, or all of these. Today, there are several hundred primate species in a spectacular array of sizes. The smallest is the size of a mouse. The largest is a gorilla. A silverback male gorilla may weigh 275 kilograms (six hundred pounds). Primates are still largely specialized for climbing about in trees, hence the opposable thumbs and toes. The so-called higher primates (meaning monkeys and apes as opposed to lemurs and lorises) have excellent vision, including color vision, making it easier to pick out ripe fruit from a green background, and (unusual for mammals) are active primarily during the day. Primates generally live in groups ranging from the size of small families to the size of small villages, something like the mole-rats. Primates also proceed through life slowly compared to other mammals. That is, they require longer care by parents, usually but not invariably by mothers alone, before they are ready to survive on their own. They take longer to reach adult size, and as adults they reproduce slowly. Perhaps it shouldn't be surprising, given this general pace of life, that primates live longer than average mammals, too.

The anatomical similarity between humans and other primates had been recognized long before Charles Darwin pointed out the obvious—that this resemblance indicates shared ancestry. Early anatomists also noticed that among primates we were most similar to the great apes—chimpanzees, gorillas, and orangutans. Darwin guessed that among the apes, chimpanzees would turn out to be our closest relative. As he was about most things, Darwin was right about that, too. How closely related are we to other primates?

The best way to assess evolutionary relationships these days is by comparing similarity in the sequence of DNA *nucleotides*, that make up a species' genome. Whole genomes—that is, the entirety of a species' DNA—have now been sequenced for all the great apes and many monkeys. From those DNA sequences, we find that among genes that we share with other great apes, which is the overwhelming majority of them, humans differ from chimpanzees at only 1.2 percent of nucleotides. For about 30 percent of these genes, our DNA sequence is identical to that of chimpanzees. Putting those numbers in perspective, we humans differ among ourselves at about 0.1 percent of our nucleotides, a twelfth of our difference with chimpanzees. Gorillas are our next closest relative, differing from us in 1.6 percent of the nucleotides in the same genes. Finishing out our genetic relationship to the other great apes, orangutans differ at 3.1 percent of nucleotides. Moving to more distant mammalian relatives, we are about 7 percent different from a standard-issue African monkey and 15 percent different from a mouse.

THE MEANING OF BRAIN SIZE

Before we discuss the longevity of some primate species, let's ponder the meaning of brain size a bit more deeply. Thinking about the brain as one coherent entity seems almost medieval today. Even in the nineteenth century, we knew that not all parts of the brain were created equal. Different brain regions are used for different things, something we discovered by observing the effects of strokes or traumatic damage to different parts of the brain. There are regions devoted to speech, vision, touch, smell, movement, planning, reasoning, and memory as well as to housekeeping chores like keeping us breathing, our hormones flowing, and our hearts beating without conscious effort. Among animals, some brain regions may expand over evolutionary time, and others may shrink, depending on the impact of those regions on the reproduction and survival of particular species. Evolution molds brain size and shape as it molds most things. For instance, the naked mole-rat spends its life in the dark, so vision holds little importance for its reproduction or survival. By contrast, touch, particularly around the face (remember, it digs and forages head first), is very important. Consequently, millions of years of evolution have greatly expanded the part of its brain devoted to touch, particularly touch around its face and in its incisors, and shrunk its region devoted to sight.

So a reasonable question is whether the size of certain brain regions rather than the whole brain might be most associated with longevity. Human brains are distinguished by their large cortex. It is so large that it sprawls over the rest of the brain like a mushroom cap sprawls over its stalk. The cortex contains regions specialized for each of our senses and for language, and its front—the so-called prefrontal cortex—is what is usually thought of as our thinking center.

Another brain region also deserves some attention—the cerebellum. From 1850 until recently, the cerebellum (its name meaning "little brain" in Latin) has gotten little respect. It looks pretty much the same in all species, bulging out behind or beneath the cortex, changing in size in concert with the overall size of the brain. In humans, it is about the size of a baseball, pleated in layers almost like accordion bellows. For most of its history, the cerebellum was thought to be involved only in coordinating movement. That should have been surprising given that it has more neurons (the information-transmitting cells) than the rest of the brain together. If nothing else, you might have thought enough it would be enough for it to shed its "little brain" reputation, but that has only recently begun to happen. Sophisticated brain imaging techniques now are suggesting that in addition to coordinating movement, the cerebellum may play an important role in processing and organizing the flood of sensory information we receive every instant. It may also function in short-term memory, emotion, attention, higher thinking processes, and the ability to plan and schedule tasks. All of these are functions that are traditionally associated only with other parts of the brain. So cerebellar size and number of neurons could potentially be involved in longevity, too.

How do our overall brain size as well as our cortex and cerebellum stack up with the cortices and cerebella of other species long-lived in absolute number of years or long-lived for their body size?

Bats are the longest-lived mammals for their body size, as we have seen, but their brains are not large for their size. They are close to average or even a bit smaller than average for other same-size mammals. If you consider that minimizing weight is important for the powered flight, this should not be surprising. Even bats' genomes have been miniaturized. Bat genomes are about a third the size of other mammals. Despite its average size, though, is there anything about the bat cortex or cerebellum that might be exceptional?

Certainly not the cortex size. The bat cortex is small even for the size of their brain. Their cerebellum, however, is oversized, at least as a percentage of the total brain. If the weight of the brain as a whole or of its different regions seems like a crude way to assess function, it is. We can now go one better. We can now count the neurons in various brain regions thanks to a clever technique developed by neuroscientist and evolutionary biologist Suzana Herculano-Houzel in the early 2000s. It can't be done on living brains inside living skulls, but she has counted the cells in many of the bottled brains of deceased animals that abound in biologists' cabinets around the world.[2]

Her work on neuron number quickly solved one mystery, which is why absolute brain size may tell us little about brain function, including general intelligence, assuming there is such a thing. It turns out that brain weight and volume, the things that have been measured for several centuries, are poor predictors of neuron number because neuron size can vary enormously. For instance, because their neurons are larger, brown bears, with a three times larger cortex than a golden retriever dog, have fewer neurons than the dog in that cortex. Similarly, elephants and any number of whales and dolphins have larger brains than humans do. An African elephant's brain, for instance, is three times the size of ours, with three times the total number of neurons. So why don't elephants, whales, and dolphins have anywhere close to the intelligence that we have? As any real estate agent might explain, location, location, location. Humans turn out to have three times as many neurons in their cortex, the thinking part of the brain, as do elephants in their much larger cortex. Elephants' cortical neurons are simply larger than ours. In fact, we have many more neurons in our cortex than any other species measured so far, although, to be fair, we still know little about whale and dolphin brains. Most elephant neurons—98 percent of them—are packed away in the cerebellum and are doing what? We are not sure, but Herculano-Houzel believes that it may have to do with the complexity of manipulating that hundred-kilogram (220-pound) trunk. I suspect, given emerging knowledge about the cerebellum, there may be more to it than that.

Let's pause here and emphasize that neuron number should not be assumed to tell the whole story when understanding animal intelligence. Intelligence itself is devilishly difficult to understand, much less compare among species. Are dogs smarter than cats? That depends on how you define intelligence. Chaser, a Border collie, was trained to recognize a thousand items by name, and dogs can be taught to do any number of tricks. Cats are famous for not learning tricks or following commands. On the other hand, if you released a dog like Chaser into the wild, how do you think he would fare compared to an average house cat? It may be that the wiring of a brain is as important (or even more important) for understanding specific kinds of intelligence as is neuron number. On the other hand, we can count neurons, whereas we know precious little about comparative brain wiring.

So perhaps it isn't surprising that neuron number does not help explain bat longevity. As noted, the bat brain has a smaller cortex and larger cerebellum than you would expect from its overall size. In terms of neuron number, bats are similarly poorly endowed in their cortex. Even in their cerebellum, they are a pretty typical mammal. Despite the size of their brain, it still allows bats to do some remarkable things, reemphasizing that neuron number is not the whole answer to brain function. Bats, for instance, forage dozens of miles each night and have no trouble finding their roost again in the dark. Female bats even remember, in a roost that may contain several million bats, where they parked their pups. Clearly, there are limits to what we can infer from neuron number. In any case, if brain size or neuron number has a role in longevity, it does seem, from my analysis, to be confined to primates.[3] Understanding why brains are so important among primates and how brain size and longevity vary among different primate species might help us answer that question. Let start by looking at our closest relative.

CHIMPANZEES

Chimpanzees have the biggest brains and live the longest of any primates. The biggest-brained and longest-lived chimpanzee is us, humans. All right, not everyone would consider us chimpanzees, but to a zoologist, that is exactly what we are. To my human-biased eye, chimpanzees and gorillas appear more alike than humans and chimpanzees, but that is clearly not the case, as we saw when comparing genome similarity. Humans, however, are unique enough to deserve our own chapter. So more about that naked ape later.

There are two other living chimpanzee species, the so-called common chimpanzee (*Pan troglodytes*), which I will call just the chimpanzee from now on, and the bonobo (*Pan paniscus*). Both live in Central and West Africa, geographically separated by the Congo River. They diverged from a common ancestor less than a million years ago. We shared a common ancestor with them both about 6 million years ago.

Average chimpanzees of either species are smaller than average humans, but there is plenty of size overlap. Males are the larger sex in both species. Average weights for wild chimpanzees are around fifty-five kilograms (120 pounds) for males and thirty-nine kilograms (eighty-five pounds) for females, although, as with humans, there is plenty of variation around these averages. Bonobos, with their more slender build, are a bit smaller. Both species spend much of their time in trees and sleep at night in arboreal nests. They move easily on the ground as well, typically walking and running on all fours, using their knuckles. If sufficiently motivated, they can walk on two feet, although they do it clumsily. Like most primates, they eat mainly fruit, although they eat many other things, including other animals, too.

Both species live in complex "fission-fusion" social groups—meaning that within their larger colony, which may contain a hundred or more individuals, there are many types and sizes of temporary groups. A Martian ethologist would likely say that humans also live in fission-fusion social groups—at-home groups, work groups, football game groups.

Closely related chimpanzee males form the stable core of their societies. Females emigrate into new groups when they come of age. There is a well-defined, linear dominance hierarchy among the males, who settle individual or group disputes with physical, sometimes lethal, violence.

Bonobos are famously different. Females are the dominant sex. Disputes and most other social interactions are settled by having sex. Actually, bonobos seem to have sex, in the form of rubbing genitals, at all times with all comers of either sex, whether there is anything to disagree about or not.

One of the great mysteries of primate behavior is how two species that shared a common ancestor so recently and are so similar physically could differ so drastically in their social behavior.

Both species have brains about a third the size of ours. Their cortices also contain about a third as many neurons as ours. This is plenty of cortex to make them devilishly clever, as anyone who has worked around them can attest. Chimpanzees were unpopular among trainers in the Hollywood film business when I worked there because they are smart enough to be deceptive about their intentions and could embarrass you on a movie set. They are smart enough to plan ahead, solve problems with thought rather than trial and error, use a variety of tools, and even learn to communicate their wants to humans via sign language. In a short-term memory test, in which the numbers 1 to 9 briefly flash simultaneously at random locations on a computer screen, a chimpanzee named Ayumu could remember the location of each number and tap those locations in numerical order far better than a human memory champion. You might say they are smart enough to learn to ride a bicycle but not smart enough to invent one.

Wild chimpanzees, like wild elephants, reach early life developmental milestones a bit more quickly but not spectacularly differently than humans do. First menstruation in female chimps occurs at ten to eleven years old in wild populations, with first birth coming along two to three years later. These events are accelerated by a few years in captive chimps, with their abundant high-quality food that requires no work to obtain. This allows them to grow faster and mature earlier.

Now for the key question. How long do chimpanzees live? As we know a lot more about chimpanzee than bonobo longevity in the wild, I focus only on that species from here on. Chimpanzee longevity, whether in the wild or in the zoo, turns out to be a complicated story that should probably begin with a story of a specific chimpanzee named Cheeta, animal star of the 1932 and several subsequent Tarzan movies also starring Johnny Weissmuller.

I came across Cheeta's story in a discarded newspaper. That story reported that Cheeta had recently celebrated her seventy-second birthday. As I've mentioned before, in a previous life I trained large cats for the movies. One thing I took away from that experience was that anything you read or hear in the news about a movie animal is probably a lie. On the other hand, the oldest chimp the aging research community knew about at that time had lived for only fifty-nine years, so if Cheeta was the real deal, it was a scientifically interesting discovery, suggesting that our closest animal relative lives substantially longer than we thought.

To investigate, I phoned Dan Westfall, mentioned in the story as Cheeta's owner, at his animal sanctuary in Palm Springs, California, and asked how he knew Cheeta's age. Westfall told me he had inherited Cheeta from his uncle, animal trainer Tony Gentry, who had personally purchased Cheeta as a juvenile in Liberia in 1932, flying back with him to the United States in time for the first Tarzan movie. This one transfer, from one family member to another, seemed to ensure that Cheeta's exceptional longevity was not a case of mistaken identity. Westfall also told me that Cheeta had been retired after a role in the 1967 movie *Dr. Doolittle* starring Rex Harrison. That sounded odd because adult male chimps (Cheeta would have been in his late thirties by then) are unpredictable, aggressive, and much, much stronger than humans—much too dangerous to turn loose on a movie set, in other words. Juvenile chimps are the movie stars, which is why many people don't realize how big full-grown chimps are. Cheeta, for instance, reportedly weighed about seventy-two kilograms (160 pounds).

In the service of science and wanting to confirm Cheeta's age, I bit the bullet, bought all the Weissmuller Tarzan movies, and watched each chimp scene in stop action—repeatedly. I want full credit for this because these are some of the most dreadful movies ever committed. I paid particular attention to the chimps' ear whorls, which change less with age than other physical features. Of the many different animals of both sexes called Cheeta in these movies, one from the 1932 version looked to me like its ear whorls *could* have matched the surviving Cheeta. I never saw an adult chimp in *Dr. Doolittle*, only a small juvenile, but that film was so stupifyingly awful that I couldn't force myself sit through it twice to make sure. I assumed Cheeta must have flashed by on the screen when my attention wandered. Who would make up a story so easily disproved?

And so I began telling all my scientific colleagues that we had been grossly underestimating how similar to us chimpanzees were in terms of aging. I gave talk after talk at scientific conference after scientific conference, making this point. In the meantime, Cheeta celebrated his seventy-third, seventy-fourth, and seventy-fifth birthdays, each year setting a new longevity record that was acknowledged in *The Guinness Book of Records*.

And then in 2008, an article appeared in the *Washington Post* called "Lie of the Jungle" by R. D. Rosen. Rosen had gone to California to do a book on this extraordinary chimp, but the book project fell apart rapidly. He discovered that there were no commercial transatlantic flights in 1932, so Gentry couldn't have flown from Liberia to the United States with or without a chimp. And, yes, no matter how carefully you watch it, there are no adult chimps in *Dr. Doolittle*. Rosen contacted some other movie animal trainers—I'd even worked with one of them—who knew Gentry back in the day, and they unanimously agreed that he had never worked on a Tarzan movie. They did recall, however, that he had been given a juvenile chimp when a famous Southern California amusement park closed in 1967. So instead of being in his mid-seventies, Cheeta, it turned out, was probably in his mid-forties, a geriatric but not exceptional chimpanzee age and quite consistent with what we previously had known about this species. I drew two lessons about this incident. One, scientists—at least this scientist—make terrible investigative journalists. Two, in Hollywood, even the chimps lie about their age. I mention my embarrassing story about Cheeta to stress once again how careful even skeptical scientists have to be about verifying longevity records, especially among captive animals where a profit motive may lurk. Did I mention, you could purchase a signed painting done by this famous movie chimp?

There are several limitations on our knowledge about wild chimp longevity. First is the length of time individual wild chimps have been observed compared to how long they live. The longest-running study of wild chimps is that of Jane Goodall, who began her famous Gombe chimpanzee project in 1960, when Tanzania was still known as Tanganyika. Over the next decades, several other long-term chimpanzee studies began. So at most, we have been tracking chimpanzees in the wild for about sixty years and in other populations considerably less. It is also important to recall that over this time, chimpanzees, like elephants, have been under siege by encroaching human populations, habitat destruction, hunting, and in the case of chimpanzees, infectious diseases. Since Goodall began her study, the Gombe population has shrunk from about 150 to ninety animals.

Because of our close genetic relationship, chimpanzees suffer from pretty much any infectious disease that we do and vice versa. Recall that AIDS is caused by a virus that spilled over from chimpanzees into humans. About one in ten Gombe chimps are infected by SIV, the chimpanzee virus that mutated in humans into HIV, and SIV-infected chimps are more than ten times more likely to die at any age than uninfected chimps. The Gombe chimps also suffered a polio outbreak in 1966, and flu-like outbreaks occur about once per decade. About half of all known chimpanzee deaths in Gombe are due to infectious diseases. I mention these things because this population of chimps, although the longest-studied, seem to be somewhat shorter-lived than other, less encroached upon, populations. The oldest chimp ever recorded at Gombe was a female named Sparrow. She was first spotted one September day in 1971, gave birth to her first baby about two years later, and disappeared, presumably because she died, some forty-four years after her original appearance. Clearly aged in her later years, she did not give birth during the last fourteen years of her life. Given what is known of other females at Gombe, Anne Pusey (who now directs the Jane Goodall Institute Research Center, has been involved in that study since 1970, and told me about Sparrow) judges that she was about fifty-seven years old when she died, give or take a year or two.

Sparrow's story points out one other wrinkle about determining how long female chimps live. In group-living mammals, one sex typically remains in their birth group throughout life, and the other sex sets off into the wide world to become an immigrant elsewhere. Emigration typically happens at about the time they are ready to starting reproducing, which is evolution's way of preventing close inbreeding, such as mothers with sons, fathers with daughters, or siblings with one another. In most mammals, males are the emigrating sex, but in chimpanzees, it is females. What this means for determining the age of wild chimps is that females were usually born in a different group than where they live their adult lives. So female birth dates are typically estimates. However, as we know the approximate age at which wild chimpanzees begin to reproduce, these estimates are actually quite good.

In populations where chimps have been studied for less time, the oldest chimp ages have to be estimated from a variety of other clues, such as appearance, behavior, reproductive pattern, and so on. From these studies done in less disturbed populations, it appears that wild female chimpanzees have a life expectancy at birth of around thirty years, maybe as many as half make it to age forty-five, and a select handful make it into their sixties, possibly even their late sixties. One wild Ugandan chimp still alive at this writing is estimated to be sixty-three, another in a less disturbed population is thought by researchers to be as much as sixty-nine years old, a longevity quotient of 3.3. Although Cheeta's age turned out to be bogus, it may not be as unrealistic as I originally thought.

Except for one thing. Cheeta—at least, the version of movie Cheeta, who supposedly reached the age of seventy or so—was a male. Like humans but even more so, male chimpanzees are shorter-lived than females both in the wild and captivity. Wild males have a 20 percent to 30 percent lower life expectancy than females, and the oldest estimate I can locate for a wild male is fifty-seven years. These life expectancies, I want to point out once again, would not be exceptionally short for some human hunter-gatherers. The extremes are very different though. The longest-lived humans, even those living in a state of nature, survive considerably beyond their late sixties, like the chimpanzees. This again shows the limitations of life expectancy at birth as an adequate description of longevity patterns.

So now that we know more about wild chimpanzee longevity than I knew when I was chasing down Cheeta's real age, how does the fifty-nine-year captive longevity maximum that researchers in the aging field had been using look? Not so good, if chimps live about that long and maybe up to a decade longer in the wild. Chimpanzees in captivity have their challenges, not the least of which is that it is difficult to keep these highly social animals in appropriately assembled social groups and keep them free of human diseases.

To find out the latest on captive chimps, I contacted Steve Ross at Chicago's Lincoln Park Zoo. Ross keeps the North American chimpanzee studbook, which maintains birth, death, and reproduction records for all the zoo chimpanzees in North America. Ironically, most of the oldest zoo chimps had some of the same problems with precise birth records that we have with wild chimps. Namely, the oldest ones were born in the wild and came to the zoos as juveniles or adolescents many years ago, so their age at arrival was estimated, and their arrival documentation is from a precomputer era when zoo records were less carefully maintained than they are today. Still, if they were fairly young upon arriving at a zoo with rigorous record keeping, such estimates could be pretty accurate. The oldest chimp in an American zoo whose age we can verify without question is named Wenka. She is one of the few older females who was born and lived her whole life in the same facility, so we can be certain of her age. At the time of this writing, she is still alive at sixty-seven years old. In addition to Wenka, a recent report from Japan describes a *male* chimp named Jhonny, who lived at the Kobe Oji Zoo for sixty-six years and was estimated to be two years old when he arrived from the wild.[4] Estimating the age of a two-year-old chimp is not difficult, so there is little reason to question his estimated life of sixty-eight years. Still, even Jhonny lived only about as long or possibly a bit longer than several estimates for wild chimpanzees, suggesting to me that we may not have reached the limit of captive chimpanzee longevity yet. By the way, the captive chimpanzees in Japan, according to this recent paper, have a life expectancy at birth of 28.3 years, about the same as reported for wild chimpanzees. It is hard to believe that chimpanzees should not be living longer in captivity than they do in the wild. Maybe we don't know the best way to care for them yet, or maybe, like elephants, they simply do not take well to the stresses of captivity.

One other female, born in the wild and captured when still young, may tell us more about how long chimpanzees can potentially live in a zoo if—and it's a big if—you can believe her story.

Let's be clear: the story of Little Mama (figure 10.1) should be taken with a sizable grain of salt. It could be true, but it could also be a case of mixed-up identity. We can't be sure. We can be sure of the following. She was born in the wild and arrived in the United States in the 1940s, probably as a teenager. She is officially listed as number 110 in the *North*

Figure 10.1
Little Mama, claimed to be the oldest chimpanzee ever. There are many reasons to be skeptical about Little Mama's reputed age of seventy-nine years, nine months, although she certainly looks ancient enough here, a year before her death. If true, she would have lived more than a decade longer than any chimpanzee with a validated birth and death record.
Source: Photo by Rhona Wise/AFP/GETTY.

American Regional Chimpanzee Studbook. Her early years go a bit fuzzy, always concerning from a record-keeping point of view. There are stories that she was a private pet for a time. She may have performed as an ice skater, possibly in the famous Ice Capades. Uncontroversially, she showed up at Lion Country Safari, a drive-through zoo near West Palm Beach, Florida, when that facility opened in 1967, where she remained for the next fifty years until her death in 2017. The official zoo story is that Jane Goodall spent time with Little Mama in 1972 and estimated her birth date as 1938. Estimates of the age of old chimps are highly error prone even if made by a famous primatologist, but zoo officials eagerly accepted it. Moreover, they decided that Valentine's Day should be considered her official birthday and in her later years held an annual, highly publicized Little Mama birthday celebration. Is this starting to sound familiar?

Incidentally, I contacted Jane Goodall about Cheeta as well because his owner, Dan Gentry, had told me that Goodall could vouch for Cheeta's age. Her response as I remember it was something like, "Wow, if that animal is still alive, he would be ancient." Of course, it wasn't the same animal, as we now know.

Goodall apparently returned to visit Little Mama in 2015, and the zoo PR folks released some treacly verbiage about how Little Mama greeted her as an old friend by making "smoochy noises" and patting Jane's hair. Little Mama received a nice obituary in the *Palm Beach Post* on November 14, 2017, seventy-nine years, nine months after her guesstimated birth. This would give her a longevity quotient of about 3.8.

There are many reasons to be skeptical of Little Mama's great age claim. There is her complicated history, during which there were many opportunities for mixing up identities. More problematic to me is her much greater age than any other chimpanzee on record. No other chimp, captive or wild, is known or even estimated to have reached age seventy, yet here is one supposed nearly eighty? If she were a human, it would be like finding a single 140-year-old in the local nursing home. On the other hand, if chimpanzees in the wild can live into their sixties, I suppose it is conceivably possible that one chimpanzee could live ten to fifteen years longer than any other wild or zoo chimpanzee. If valid, that zoo record would tie Asian elephants for second place in record longevity among terrestrial mammals. More likely, particularly given that their overall life expectancy is considerably shorter than either elephant species, chimpanzees should rank number four among terrestrial mammals in absolute longevity. In terms of relative longevity (that is, LQ), chimpanzees and other primates are no bats.

ORANGUTANS (*PONGO* SPP.)

Orangutans may be the next longest-lived terrestrial mammal in absolute number of years lived. Orangutans have played various roles in the human imagination for centuries. Edgar Allen Poe, for instance, had probably never seen a live orangutan when he made one a villain in his 1841 short story "The Murders in the Rue Morgue." In that story, there was a gruesome double murder, a mother with her throat slashed and her daughter strangled, that turns out to have been done by an enraged orangutan using a stolen straight razor and its superhuman strength. It's a good thing that orangutans don't actually have maniacal personalities like this because they surprisingly often end up walking among us.

With the second-biggest brain among primates after humans and an encephalization quotient of about 2.9, orangutans have repeatedly demonstrated their high level of intelligence by escaping from their zoo enclosures. They have escaped by scaling seemingly unscalable walls, by disassembling key enclosure parts, and by fashioning tools to jimmy locks. An orangutan named Fu Manchu at the Omaha Zoo not only fashioned a tool with which to jimmy the lock on his enclosure but also hid the tool from his keepers. As an infant, a particularly famous orangutan escape artist named Ken Allen in the San Diego Zoo repeatedly escaped from his cage by unbolting its wire top, spending nights exploring the zoo nursery, and then bolting himself back inside before zoo personnel arrived in the morning. As an adult, he escaped from his enclosure multiple times, even once finding a crowbar and tossing it to another orangutan so that she could pry open a window to let him out. As I say, fortunately orangutans are not particularly aggressive, and I know of no cases in which they have injured zoo visitors during their escapes.

Orangutans are distinctive among the living great apes in that they do not live in complex social groups. In fact, they do not live in groups at all. They are largely solitary, ranging over vast distances in often swampy tropical forests in Indonesia, and this makes them excruciatingly difficult to study. In these most arboreal of apes, juveniles remain close to their mothers for years, which is as close as they come to living in groups. Sometimes a few animals may happen to congregate at a fruit-laden tree, but for the most part, their time is spent alone, clambering slowly through the forest canopy looking for that next fruiting tree. The emphasis here is on "slowly." They are the slowest of the apes in several senses. They move slowly through the trees, as you might expect from an arboreal animal large enough to be seriously injured or killed by a fall. Also, as you might guess from watching their stately progress through the forest canopy, they have an exceptionally slow metabolism—in fact, the slowest for their size of any advanced mammal except for sloths.[5] Sloths move as if they are swimming through honey, as you might have heard. Orangutans not only move slowly; they proceed leisurely through life's stages. They grow slowly, take their time to start reproducing, and reproduce slowly once they start. All this would seem to suggest that they should age slowly as well.

Today, orangutans are found only on the tropical islands of Sumatra and Borneo, and they are under heavy siege by destruction of their forest habitat. There are three species—two on Borneo and one on Sumatra—but they are similar enough that until fairly recently, they were considered a single species. So I here ignore subtle differences among them. Once, however, they roamed the South and Southeast Asian mainland, reaching as far as southern China. Orangutans are roughly human size, though with much shorter legs and longer arms, useful for clambering about in trees. Their arms actually reach to their ankles, which would be useful for tying their shoes if they wore them. They move clumsily when on the ground but may stand upright on tree limbs in a very Tarzan-like pose.

I say they are roughly the size of a human, but I should add that orangutan size is easy to mischaracterize. Those orangutans you see in movies, such as in Clint Eastwood's woebegone 1978 film *Every Which Way But Loose*, are small because, like movie chimps, they are youngsters, adults being too strong and independent-minded to risk injury to actors. In fact, by the time of Eastwood's even worse 1980 sequel, *Every Which Way You Can*, a different orangutan was used. The original one, named Manis, was by then too big to safely use. On the other hand, zoo orangutans—because of their slow metabolism, the abundant, high-quality food that they don't need to work to get, and the fact that males continue to grow throughout their adult life—reach enormous sizes that wild orangutans never approach. The largest males in zoos have approached two hundred kilograms (440 pounds) in weight, more than twice the size of males in the wild.

In nature, females reach their adult size of thirty-five to forty kilograms (seventy-five to ninety pounds) and have their first baby at about age fifteen. Typically, it will take another eight or nine years before they follow up with baby number two. Notice that both of these numbers—age to maturity and time between births—are several years longer than wild chimpanzees take to reach the same milestones. Because of the difficulties of finding and keeping track of solitary wild orangutans through the rainforest, their longevity in nature is poorly known compared with chimpanzees, even though one study in Sumatra has lasted more than thirty years. Most of what we know about their later lives and longevity in the wild are crude estimates. Our most solid knowledge comes from zoos and wildlife rescue sanctuaries, but we don't know how well that may represent life in the wild. In zoos at least, females stop reproducing in their late twenties. However, as zoo orangutans survive no better and possibly a bit worse than animals in the wild, this age could easily not represent when reproduction halts in nature. In zoos, female life expectancy at birth is about twenty-five years. If they survive to adulthood, they average about thirty-five years.[6] The longest-lived captive female on record was estimated to be fifty-eight years when she died at the Philadelphia Zoo in 1976.[7] I say "estimated" with particular emphasis for reasons I'll explain shortly. Other females have been reported to live into their early to mid-fifties in zoos, however, so I don't completely discount this fifty-eight-year record. The oldest *wild* female was estimated to be somewhere between forty and fifty-three years old, the breadth of that estimate saying all that needs to be said about the difficulty of estimating age in wild orangutans.[8] So let's agree that female orangs live to, say, at least their middle fifties in zoos and maybe close to that in the wild, too, which would make them number five—behind humans, two elephant species, and chimpanzees—in the absolute longevity rankings of terrestrial mammals. Let's hold off on their longevity quotient for now.

Male orangutans need to be considered separately due to several intriguing peculiarities of adult lives, some of which remind me of male elephants.

First, although they appear similar to females in size and shape when both become capable of reproducing in their midteens, males continue to grow throughout life, so that by the time they reach full physical maturity, they are about twice the size of females, some seventy-five to ninety kilograms (165 to two hundred pounds). Because of continued growth, older male orangutans (like older male elephants) are larger than younger ones.

Here is where they become unique and most interesting. Although they become physically capable of reproducing in their midteens, they may continue to resemble adolescents (that is, they continue to look a lot like females) for considerably longer—sometimes as much as a decade or two longer—while continuing to grow. Eventually, if they survive long enough, they will develop fully adult features—specifically, a large throat pouch and a special male orangutan facial feature, large cheek pads called *flanges* (figure 10.2). The flanges and throat pouch are signs of dominance. As flanges develop, males grow longer, darker hair and also put on muscle. Although males occupy large, overlapping ranges, the biggest, baddest male in the area continually announces his presence with special loud calls produced in that throat pouch. Those calls reverberate through the forest for a kilometer or more. The presence of a flanged, dominant male in an area appears to suppress the development of flanges and throat pouches of other males. So smaller, younger, although reproductively capable males have their full male development arrested. If two flanged males happen to meet at, say, a fruiting fig tree, it frequently leads to violence. If the local flanged male disappears, a previously beardless, flangeless male may quickly develop his own flanges and throat pouch. Fertile females tend to avoid unflanged males who will pester them to mate, but they are attracted to flanged males, because—well, how could anyone resist those flanges and that voice?

How long do male orangutans live? In the wild—again, because of their long life and the difficulties of monitoring them as they range alone far and wide in the forest—there are only crude estimates. So far, the oldest was guessed to be somewhere between forty-seven and fifty-eight years old, and several others died at estimated ages just a few years younger. Official zoo records of male orangutans show a life expectancy of about twenty years at birth or twenty-eight years if they survive to a reproductive age.

Figure 10.2
Unflanged and flanged male Tapanuli orangutans. The Tapanuli species (*Pongo tapanuliensis*) shown here lives on Sumatra. It has only been recently described and is highly endangered. Note the large cheek pouches and more extensive beard of the flanged male. The beard hides a large throat pouch used to produce loud calls, attractive to fertile females, that may resound through the forest for a kilometer or more.
Source: Photos by Tim Laman.

Notice, these numbers are a few years less than those of females, something similar to other great apes. Interestingly, estimated life expectancies from natural populations suggest that males live a bit longer. As for the longest-lived individuals, official records list a male who died at a presumptive age of fifty-nine years in 1977. If you remember my skepticism about the oldest recorded—fifty-eight-year-old—female, I'll now explain why. By a remarkable coincidence, the oldest male and oldest female ever recorded were acquired at the same time (1928) from the same place (Havana, Cuba) and shipped to the same zoo (Philadelphia) on the same date (May 1, 1931) by the same person (Mrs. Abreu), who claims they were wild-born in the same year (1918).[9] Certainly, remarkable coincidences do exist, and maybe this is one of them. But with all the modern improvements in the care of zoo animals (from which orangutans, in particular, seem to have benefited)[10] and the advent of computerized zoo record keeping, the fact that the oldest orangutans ever recorded came from so long ago when zoos of the world were less adept than now at keeping them alive has my nonsense detector flashing. There are more recent zoo records of orangutans living to their early fifties, close to the longevity records estimated in nature. Orangutans clearly seem capable of living into their fifties, possibly their late fifties, but they are no chimpanzees, human or otherwise.

What about the orangutan's longevity quotient? How should it be calculated when the sexes differ dramatically in size? Do we have one LQ for females and another for males? Actually, that would not make sense. Remember the reason we use LQ is to account for *species* differences in longevity. Larger species generally live longer than smaller ones. That does not mean that larger individuals *within a species* live longer than smaller ones. In fact, it is often quite the opposite. Smaller individuals often live longer. So species need to have a single LQ, and averaging the size of the sexes and accepting that fifty-nine-year-old longevity record, even if it may be a squishy record, we come up with an LQ for orangutans of 2.6. Although they are the third-longest-lived primate species (after humans and nonhuman chimpanzees) in absolute number of years lived, they have a pretty average primate longevity for their size.

CAPUCHIN MONKEYS (*CEBUS CAPUCINUS*)

If you accept the accuracy of the record, the largest longevity quotient for any primate, including the naked chimpanzee, belongs to a New World monkey that is considerably smaller than any ape.

Primates originated in Africa and were entirely absent from the New World until 30 million to 40 million years ago, when at least one monkey of each sex rafted on floating vegetation, perhaps dislodged during a storm, across the Atlantic Ocean to South America. The Atlantic was considerably narrower in those days, so this rafting adventure, while pretty remarkable, is not as extraordinary as it might seem. Also, there may have been a series of mid-Atlantic islands in those days, so the journey could have been done by multigenerational hopscotching. In any case, those original immigrants by now have diversified into more than 150 monkey species, ranging in size from the 120-gram (4.2-ounce) pygmy marmoset (*Cebuella pygmaeus*) to the twelve-kilogram (twenty-six-pound) woolly spider monkey or, as it is called locally, the muriqui (*Brachyteles arachnoides*). No apes, aside from humans or even large monkeys such as baboons, exist in the New World. Ours is a small-primate continent.

Small primates need to avoid ground-based dangers as much as possible, so they spend virtually all of their time in trees, where they eat mostly fruit but also flowers, leaves, buds, bark, sap, and small animals. Some New World monkeys evolved a new primate feature, a prehensile or grasping tail, which helps them navigate the forest canopy. The most interesting group of New World monkeys from a longevity perspective is the capuchin or organ grinder monkey.

Organ grinders were the buskers—that is, the musical panhandlers—of their day. Coming to the United States with a wave of Italian immigration in the latter half of the nineteenth century, murdering silence with their hand-cranked street organs, they found it increased profits to have a costumed pet monkey offering up the donation cap or cup to onlookers. Small enough (about the size of a small cat) not to be life-threatening to people, smart enough to learn tricks, dexterous enough to hold a cap or cup, tolerant enough to be dressed in outlandish costumes, capuchins not only worked with organ grinders, but they also became popular exotic pets for people best described in a book by my wife that she called *Normal People Don't Own Monkeys*. The movies loved their cuteness and trainability as well. Capuchins have had roles in dozens of movies, my favorite being the red-vested, pet monkey spy named Sieg Heil in the movie *Raiders of the Lost Ark*. Why a South American monkey would be a pet in Egypt, which has plenty of its own native monkeys, is a question that only a confused casting director might be able to answer. One thing readers should know about pet monkeys, which are now thankfully outlawed in many places, is that they will bite you. It isn't a question of where or when or under what circumstances. As inevitably as the sun will rise or the tides ebb and flow, a pet monkey will bite its owner, its owner's children, and its owner's friends. However, in the wild, where they should be, capuchin monkeys are fascinating.

They live in groups of one to two dozen and tend to hang within eye- or earshot of one another as they prowl the forest looking for fruiting trees. With their five-limbed agility, they can travel through the branches much faster than you can keep up with them on the ground. Adult males have a clear dominance hierarchy for access to food and mates, but they cooperate to defend the group's territory. Females tend to be close relatives and cooperate in childcare, even nursing one another's babies. It takes some six to seven years for both sexes to reach adulthood in the wild (several years less in captivity), at which point the males leave, often as a group, to seek other groups from which they will attempt to oust the resident males and start their own families.

Capuchins have big brains for their size (their encephalization quotient is greater than 3), which is maybe why they are so trainable and also no doubt why they eventually get tired of being told what to do and bite. Their intelligence and their manual dexterity make it less surprising that they have managed to discover how to use a variety of tools to break open nuts. Capuchin intelligence is astonishing, although not as astonishing as that of chimpanzees or orangutans. Primatologist Frans de Waal characterizes the difference by saying that apes solve problems by thinking about them and capuchins do so by trial and error.

In the wild, where they have been studied since the mid-1980s in Costa Rica, they do not seem overly exceptional in how long they live. Males are the much shorter-lived sex. More than half die before they reach adulthood, and only a handful reach their midteens. Females, on the other hand, commonly live into their teens, with the longest-lived reaching at least their mid-twenties.[11] From these numbers, they do not seem to be even the longest-lived New World primates. The larger muriqui, or woolly spider monkey, commonly reaches its mid-twenties in the wild, and the longest-lived reach their mid-thirties. Here is the intriguing thing about capuchins, though. In zoos, they reach their forties and even fifties. The longest-lived on record is a wild-born *male*, captured as a baby, who *maybe* lived fifty-four years.[12] That is a longevity quotient of 4.3—greater than any ape, possibly including us. I'll explain the emphatic "maybe" shortly.

There is a pattern at work here that deserves pointing out. Among mammals, at least, the difference between their longevity in nature and their longevity in zoos seems to depend on size. Nothing could emphasize the hazards of the real world to small mammals much better than this. Remember, mice, on average, live only three to four months in the wild but up to three years in captivity. Naked mole-rat workers live more than ten times as long in captivity as in the wild, and queens live more than twice as long. Capuchins appear to live two to three times longer in zoos as in the wild, but chimpanzees and orangutans do not seem to survive much better in zoos than they do in nature, and elephants survive a bit worse.

Now about that capuchin who "maybe" lived fifty-four years. The best reason to use the longest-lived individual to characterize longevity in a species is that it may represent the *potential* of a species for long life under ideal conditions. However, the downside of using longevity records is that, as we have seen, they are prone to errors and exaggeration. I suspect this is true of the chimpanzee Little Mama, largely because she was reported to live eleven years longer than the second-longest-lived chimpanzee. Something similar seems to be going on with the oldest capuchin. That animal was wild born, reportedly arrived as an infant at the Mesker Park Zoo in Evansville, Indiana, in January 1935, was transferred to a research lab in 1980, and died there in 1988, fifty-three years after arriving in the United States. Two things concern me about this record. First, it is six years longer than the second-oldest capuchin ever reported. Second, by a remarkable coincidence, the second-oldest capuchin ever reported lived forty-eight years, forty-seven of which were also at the Mesker Park Zoo. It seems remarkable and a little far-fetched to me that Evansville, Indiana, is such a haven for long-lived capuchins. The oldest capuchin that didn't spend most of its life in the Mesker Park Zoo and was not born in the wild at some estimated date (in other words, that was born and lived its entire life in the same zoo) was a female from Chicago's Brookfield Zoo who died at forty-two years, ten months of age.[13] If the real (that is, accurate) record longevity is more like the early to mid-forties, then that makes the capuchin's longevity quotient only about 3.6, not too shabby but not quite as spectacular as the naked chimpanzee that builds space ships.

WHY ARE PRIMATES SO LONG-LIVED?

Returning to our original question, why are primates so long-lived, and why does large relative brain size seem to lead to even longer life among primate species? One thing I haven't mentioned so far is their metabolic rate. Primates as a group have a slow metabolism, expending only about half as much energy on a daily basis as other mammals of their size. This may not be so surprising for orangutans, who do everything slowly, but it is also true for chimpanzees and monkeys that we typically think of as highly active.[14] So a slow metabolic rate, I would say, is one unexpected contributor to primate longevity.

What about brain size? I would make the case that brain size, as it is reflected in better memory and decision making, helps reduce environmental dangers for primates. Being arboreal—as most primates are, at least to some extent—by itself reduces their exposure to environmental dangers. Danger from floods and terrestrial predators are certainly reduced, for instance. Living in groups (again, as most primates do) provides additional safety. Living in complex social groups requires considerable judgment, something that having a bigger brain may facilitate. A bigger brain may also facilitate memory and decision making about where food might be found at certain times of the year, how to avoid predators such as snakes and birds of prey that can follow them into the tree canopy, and how to make fewer errors of judgment about navigating the dangerous three-dimensional environment high above the forest floor.

Am I making too much of the impact of primate brain size on longevity? Possibly. I may respect our big human brain more than it deserves, too. Clearly, a big brain isn't necessary to live a long time. Just ask a bat or a tortoise. On the other hand, it may not hurt, either.

What can we learn about extending our own health from understanding that of other primates? Probably not a lot because we have, after all, the longest, healthiest life of any primate. For our size, we are almost as long-lived as (or maybe longer-lived than) capuchins. No, the reason to study patterns of longevity in primates is not for the same reason we might want to understand more about bird, bat, or naked mole-rat longevity—which is to see if we can discover evolutionary longevity tricks that could help us live a longer, healthier life. The reason to study patterns of primate longevity is to learn more about our own evolutionary history.

III LONGEVITY IN THE SEA

11 URCHINS, WORMS, AND QUAHOGS

The ocean is vast. The fact that it covers 71 percent of the earth's surface does not begin to capture that vastness. More illuminating perhaps is the fact its *average* depth is almost 3,700 meters—that is, 2.3 miles, more than twice the depth of America's Grand Canyon. In its deepest regions, it is more than a mile (sixteen hundred meters) deeper than the height of the terrestrial world's highest mountain. If you combine all of the earth's fresh water—from all the earth's streams, all its great rivers (the Amazon, Nile, and Mississippi), all the small rivers, all ponds, all lagoons, all lakes (whether great like Superior, Victoria, and Baikal or small), all the snow from all the earth's mountain tops, and all the ice tied up in the world's glaciers and both polar ice caps—all this together comprises only 3 percent as much water as there is in the ocean. As living organisms occupy every part of the ocean from its surface to its deepest darkest parts, it forms 90 percent of the earth's biosphere. Yet it is the region of our biosphere about which we know least. Oceanographers estimate that only 20 percent of it has even been mapped and that more than 90 percent of its living creatures still await discovery.

The ocean is ancient, too. Formed from millions of years of rain that began falling about 4 billion years ago, after the earth's surface had cooled sufficiently from its molten beginning for water to exist in a liquid form, the earliest ocean may have covered the entire globe, preceding even the birth of continents. Almost everyone agrees it was also the cauldron from which life emerged. One prominent theory is that life first evolved around hydrothermal vents on the ocean floor. These vents are fissures in the earth's crust through which cold seawater seeps to meet the molten magma of the underlying mantle. When cold meets heat, the resulting explosive mixture gushes back through the seafloor, bringing energy, gases, and minerals from the deep earth. That mineral-rich seawater may be as hot as 400°C (750°F), much hotter than water would normally boil because of the hundreds of atmospheres of pressure in the deep ocean. Energy, minerals, and water erupting from the earth's core are three prerequisites for life.

The ocean was the only place on earth where life, in the form of microbes, existed for its first few billion years. Life's big breakthrough came when some of these organisms discovered how to escape reliance on the paltry amount of energy being released into the deep ocean from the earth's interior by capturing the sun's abundant energy from surface waters. Photosynthesis—the process by which the sun's energy converts carbon dioxide and water into (as the literal term reveals) carbohydrates—provides energy-rich food for organisms that cannot capture the sun's energy themselves. The byproduct of photosynthesis—oxygen—became the chemical stimulus essential for an opposite chemical process, respiration, which converts carbohydrates back into carbon dioxide and water and in doing so releases the sun's energy in usable packets that can power life's processes. That additional energy allowed the evolution of organisms built of many cells. Over the next half a billion years, oxygen converted the earth's atmosphere from one based largely on carbon dioxide to one in which oxygen played a major, life-giving, role.

Despite its great size, age, and long history of life, the ocean contains about only 15 percent of earth's species, if we ignore microbes. The reason for this surprising paucity of life is that the ocean in most parts is a desert.

By that, I mean that land deserts, despite abundant energy from the sun and minerals from the soil, are defined by scarcity in the third of life's key necessities—water. This is why they harbor few species compared with, say, prairies or forests. Most of the ocean is a desert because these three ingredients come together in only a few places. Most of deep ocean bottom is barren sand and sediment. Water, as you may have noticed, is abundant in the ocean, but the sun's energy is limited to near the surface. Water absorbs the sun's energy. Sunlight can power photosynthesis to a depth of about only two hundred meters (660 feet), considerably less if the water is turbid. I have scuba dived in muddy water that seemed to me as dark as a cave at thirty meters (a hundred feet). The minerals required for life are also rare in the ocean. They are mostly packed in the seafloor sediment thousands of meters below where the sun can penetrate. Two main places where abundant sun, water, and minerals come together in the ocean—the shallow continental shelves and areas where upwelling brings seafloor sediment to the surface—are where sea life is most abundant. Life is also rich and diverse around hydrothermal vents and vents of another kind called *cold seeps* on the ocean floor, but those are few and far between relative to the vastness of the ocean.

Yet the ocean, amazingly enough, is where all of the earth's longest-lived animals live. So far, we have encountered a number of species, from birds to bats to mole-rats, that are long-lived *for their body size* and metabolic rate— much longer-lived than humans, in fact, if you take these things into account. We have a lot to learn from such species about mechanisms that protect against the damage, such as damage due to oxygen radicals induced by metabolism. Yet in absolute number of years lived, the only animals we have met so far that were clearly longer-lived than humans have been the giant tortoises and pos-sibly the tuatara. Their longevity was in large part due to their ectothermy, their size, or the cool temperatures where they live. Recall that ectothermic animals have a slow metabolism compared with endothermic birds and mammals, such that a man-size crocodile eats only one-twenty-fifth as much as an actual man. Metabolism also slows with increasing size and decreasing temperature. The long life of tuataras is due to ectothermy in the cold and the safe environ-ment provided on islands. The tortoise's longer life comes from a combina-tion of ectothermy, size, safe island habitats, and a protective shell. However, get ready for takeoff because the ocean, especially in those few places where the ingredients for life are found together, is where ectothermy, cold, and safe environments coincide time and time again. That is why virtually all of the longest-lived species found in nature live in the sea.

Life is longest in cold ectothermic animals in which life churns most slowly. Virtually all ocean life is ectothermic. Virtually all ocean life lives in the cold. This is because the sun heats only a thin ribbon of surface water, which floats over the colder, denser water underneath. The deepest 90 per-cent of the ocean is eternally dark and cold, within a few degrees of freezing.

Also, it is close to impossible to be endothermic and live in the deep ocean because water is very efficient at wicking heat away from warm bodies. The colder that water is, the more efficient it becomes. Think of how you grow chilly when lying still even in a tub of tepid bathwater. Your resting body cannot produce heat as fast as the water wicks it away. Small endothermic animals, because of their large surface area relative to their body mass, cannot permanently survive in even tepid water much less cold water. There are no mouse-size birds or mammals that live in the sea. The smallest mammal to live full-time in the ocean is likely the fifteen- to forty-five-kilogram (thirty-two- to ninety-nine-pound) sea otter (*Enhydra lutris*), and it does so by having thick waterproof fur and spending most of its time floating belly up on the sea surface grooming its heat-saving fur. Even at that, sea otters have a heat-producing metabolic rate two to three times higher than a terrestrial mammal of the same size and consequently need to eat two to three times as much food. The baby of the smallest whale, the dwarf sperm whale, is about the same size and weight as a sea otter. It also spends most of its time floating on the ocean surface, where the sun can warm its back and dorsal fin. The closest thing to a fully aquatic bird is a penguin, which has waterproof feathers, thick skin, and a layer of insulating blubber under its skin to help it keep warm when in the water. Like other so-called water birds, penguins, in fact, spend most of their time on land.

So how long do these cold-water ectotherms live? Sometimes we have a firm answer, and sometimes we need to make educated guesses. As we've seen even with easily visible giant tortoises, pretty much anything that lives longer than a human poses challenges for rigorously documenting their longevity. Once you begin talking about animals that live submerged in eternal darkness hundreds or thousands of meters below the ocean surface, the challenge increases. You are limited to observational snapshots taken during brief visits to the deep. This is why we can't be sure whether the tubeworm, *Escarpia laminata*, lives only a century or two or several millennia.

TUBEWORMS

Tubeworm is a rather imprecise name for diverse marine species related to earthworms that live inside hard but flexible tubes that their bodies secrete.

Most, like *Escarpia*, do not have common names, but some of their relatives have colorful ones. My favorites are the feather duster worm, the Christmas tree worm, or my absolute favorite, the bone-eating snot flower.

The tubeworms of interest here live in the deep ocean, either around hydrothermal vents or cold seeps. Cold seeps, as their name implies, are fissures in the ocean floor through which oil and natural gases leak from subterranean pools, providing minerals and energy for their own deep-sea animal communities. They are not colder than the surrounding seawater, which at the seafloor is near freezing, but unlike hydrothermal vents, they are not warmer, either. The Gulf of Mexico fairly bristles with cold seeps and as a consequence also bristles with oil rigs tapping the liquid gold that are the seeps' source.

Tubeworms that live around hydrothermal vents live fast and die young. They are not long-lived both because the warm temperature of the seawater accelerates their growth and metabolism and because the hydrothermal vents on which they rely for energy are not long-lived. Earthquakes, volcanic eruptions, and changes in the flow of the subterranean magma shut down hydrothermal vents on a timescale of decades. Cold seeps, on the other hand, can last for millennia.

Deep-sea tubeworms live in clusters, looking almost like desert shrubs or clusters of drinking straws stuck in the seafloor. Spending their lives inside their tubes, they lack eyes, appendages, a mouth, an anus, and indeed any vestige of a digestive tract. Without a digestive tract, how do they eat, you might ask? They don't. They are fed by microbes that live inside their tissues. To nourish the themselves, those microbes use oxygen, carbon dioxide, and other gaseous nutrients absorbed from the water (through the worm's thin fringed end that protrudes from the tube top when they are feeling safe) and use minerals from the other worm end, which is buried in the sediment.

I'm not sure I can emphasize enough how little we know about *Escarpia* or other deep-water tubeworms. From their shallow-water relatives, we have a reasonable idea of what their free-swimming larvae must look like, even though no one has ever seen one. We know they are tiny and, unlike adults, have a complete digestive tract—mouth, anus, the works. We have little idea how long they spend as larvae before settling to the seafloor, how long after that they start lengthening and secreting their first tube, or how long it takes

What Is an Animal?

Some species that are technically animals lack so many features of animals as we know them that comparing them with birds, bats, or bees seems unfair and uninteresting. I am ignoring them in this book. Sponges, some of which are extraordinarily long-lived, are one of these groups. Another group is the corals. Coral reefs are colonies of individual tentacled polyps living inside hard tubes, kind of like tubeworms. Individual polyps live only several years, but because the colony may live centuries, they have achieved a reputation for exceptional longevity. I consider this similar to the case of the Norway spruce Old Tjikko, whose long-lived root system has sprouted multiple reasonably short-lived "trees." That is long life in a different sense than I mean here.

Tubeworms, though, while they may lack eyes and any feeding apparatus, do have a circulatory system, including a (slowly, but we have no idea how slowly) beating heart, and even a collection of nerve cells, which if you stretch the term to its breaking point, you could call a brain. So they are animal enough for me to include them.

their digestive tract to disappear. We have no idea how long it takes them to grow to adulthood and begin their own reproduction.

But we do have a rough idea of how long they can live, and it is a long, long time. We know this because they and their tubes continue to lengthen as they get older. By measuring tube lengths and estimating their growth rate, marine scientists can crudely estimate their age.

Exceptional longevity of at least a century or more has been documented in several tubeworm species living around cold seeps, but the longevity champion appears to be *Escarpia laminata*. To estimate how old they might be, researchers using deep-sea submersible vehicles measured the lengths of a group of *Escarpia* tubeworms on the seafloor at a depth of 2,500 meters (about 1½ miles), marked them with dye, and returned a year later to record how much they had lengthened. Like other animals that continue to grow throughout life, tubeworms grow more slowly the larger become. Estimating their age from size and a short interval of growth requires a bundle of educated guesses, assumptions, and computer simulations. Using one reasonable set of assumptions, the *average* age of a group of sixty-five-centimeter- (two-foot-) long *Escarpia* was estimated at 266 years, and the longest-lived (that is, the longest) individual within that population came in at a whopping

1½ meters (five feet) in length and an estimated seven thousand years old. Using a second set of more conservative growth assumptions, the oldest individual would have been only a whisker over a thousand years old. Under either assumption, though, this is not too shabby. Although thinking about it, sitting there inside your tube in total darkness feeling the current drift by for millennium after millennium does not seem to me like the most rewarding existence.

I do need to provide a little reality check on estimating ages from growth rates, as if the "somewhere between one thousand and seven thousand years" doesn't make it obvious enough. Such guesstimates are crude. By making a range of assumptions, it is hoped that the real ages are captured within this range. *Guesstimate*, as I'm using it, should not be thought a pejorative term. It is a simple recognition of the degree of uncertainty with which scientists sometimes have to work. In the total scheme of things, comparative growth rate is probably the least reliable estimator we have, although for some very interesting species, it is all that we have.

What might we learn from a potentially seven-thousand-year-old worm about extending human health? Is there anything notable about their exceptional longevity beyond the glacial pace of their lives? These are good questions but also questions we are unlikely to answer any time soon. Creatures living a mile and a half under water at pressure hundreds of times that of our normal atmosphere are not easy to study. Normal atmospheric pressure is lethal to them, so understanding more about them will have to wait until we build research laboratories on the deep-sea floor. However, not all long-lived ocean creatures live in the abyss.

SEA URCHINS

The red sea urchin (*Mesostrongylotus franciscanus*) lives in relatively shallow water, from just below sea surface to about a hundred meters (330 feet) below that along the west coast of North America and the northern coast of Japan. Some of the other thousand or so species of sea urchins live in even shallower water—rocky tide pools, which is where most people have seen them. Still others live in the deep ocean down to a depth of about five thousand meters (three miles). This makes it all the stranger that the shallow-water red sea urchin is the longest-lived urchin species known.

Sea urchins look like pin cushions or maybe like curled-up hedgehogs. In fact, their name—urchin—comes from an old French word for hedgehog, *herichun*. One of the interesting things about urchins is that they (and their closest relatives, sea cucumbers and sea stars) are evolutionarily more closely related to humans and other vertebrates than virtually any other animal group lacking a backbone—closer, for instance, than animals such as squid, insects, or worms (tube or otherwise).

Urchin spines can inflict painful injuries if inadvertently stepped on. In some species, the spines are venomous, potentially capable of causing much worse than a painful injury. The spines can also swivel independently of one another, something like artillery can swivel on a war ship. And they can be regenerated if lost. Urchins also have hundreds of other tiny, sharp, pinching appendages called *pedicellaria*, some of which can be launched into the water to discourage potential predators. They also have hundreds of adhesive so-called tube feet that help them move, breathe, and (at least in shallow-water species) see—although not with twenty-twenty clarity. I have watched a shallow-water species work its way to the surface of a laboratory tank when someone loomed above with its dinner in his hand. If properly motivated, urchins can race around at up to half a meter (twenty inches) per day. If all that weren't enough to make them extraordinary, they have no heart and no brain, their anus is on top, and their mouth underneath. This mouth, possibly better described as a feeding apparatus, consists of complex jaws supporting five hard, sharp teeth that work something like the grasping claw on one of those arcade games with which you try to grab kitschy prizes. This apparatus is called (I'm not kidding) *Aristotle's lantern*. The ability of Aristotle's lantern to grab, scrape, dig, pull, and tear at things has so impressed human engineers that it inspired the design of graspers for moon and Mars landers.

Growth rate and size are mostly what we have for sea urchin age determination, too. Yes, like tubeworms, they continue to grow throughout life, and also like tubeworms, their growth rate gradually slows as they age. So a similar "measure-mark-wait-remeasure" method is used to estimate age in urchins. Specifically, they are marked with an injection of the antibiotic tetracycline, which binds to calcium being incorporated into hard materials such as bones and teeth, including the growing jaws of Aristotle's lantern. When later recaptured and the jaw removed, that mark can be seen under UV light, allowing the amount of jaw growth since the time of the original injection to be measured. Combine those measurements with some growth mathematics, and you have an estimate of an urchin's age.

Red sea urchins are harvested commercially for their delicious gonads—delicious to some people, anyway. Sometimes euphemistically called sea urchin "roe," in fact, it is the gonad, the organ that produces the roe or sperm—yes, there are boy sea urchins and girl sea urchins—that some people adore. Orange or yellowish, the consistency of firm custard (if you like it) or mud (if you don't), *uni*, as it is called in Japanese, has a reputation as an aphrodisiac, which makes some kind of demented sense, I suppose. Because of its commercial value, it became useful to know how long red sea urchins live in order to manage their populations wisely. The original estimate of its longevity was seven to ten years, but using the tetracycline method on more than fifteen hundred individuals, biologist Thomas Ebert of Oregon State University calculated that they could certainly live more than a century—the largest ones, which were some nineteen centimeters (7½ inches) in diameter, perhaps even as long as two centuries. After one hundred years, the urchins grew so slowly that the number of additional years were difficult to estimate. In one population, about one in ten urchins lived a century or more.[1] In case people doubted his method, he was able to confirm it thanks to nuclear weapons testing.

HOW NUCLEAR BOMB TESTING HELPS US DETERMINE ANIMAL AGE

This will come up again, so let me explain now how one good thing— arguably the only good thing—to come out of aboveground nuclear bomb testing is the ability to calibrate the age of long-lived animals. In the 1950s and early 1960s, there were hundreds of such nuclear bomb tests, which caused the amount of a rare atmospheric carbon isotope, carbon-14, to double between the early 1950s and 1963, when a nuclear test ban treaty was signed. Atmospheric carbon-14 levels have gradually returned to pre-bomb levels since then. As carbon in the atmosphere is in the form of carbon dioxide—which plants incorporate into their tissues via photosynthesis, which animals that eat plants incorporate into their tissues, and which animals that eat the animals that eat plants also do so—by measuring carbon-14 levels in hard, long-lasting tissues such as bone, shell, or teeth, one can determine if the tissue was formed prior to, during, or after this so-called bomb pulse of carbon-14.

One other thing about carbon-14, while we are on the topic: it is the same carbon isotope that is used to date other things—much more ancient things, in fact, than the age of long-lived animals. Carbon-14 naturally occurs in the atmosphere at about one atom per trillion carbon-12 atoms. Carbon-12 is the usual carbon isotope. So this low level of carbon-14 has been incorporated into living plant and animal material for eons. Because carbon-14 decays to carbon-12 at a steady rate—half of it disappears every 5,760 years—it is easy enough with sophisticated instruments to measure the ratio of carbon-14 to carbon-12 and estimate how long ago any living or formerly living object was formed. As with any measure, there is a zone of uncertainty (what political pollsters call a *margin of error*) around these measures, but that zone of uncertainty can be calculated, too. In this way, we know for instance that the famous Shroud of Turin, believed by some to be the burial shroud of Jesus, is made of linen that grew as a flax plant sometime between 1260 and 1390 rather than two thousand years ago and that the iceman Ötzi, whose long-buried body emerged from a melting glacier in 1991, lived sometime between 5,100 and 5,400 years ago. Radiocarbon dating, so-called because carbon-14 is mildly radioactive, is reliable only to about fifty thousand years ago. Anything older than that has so little carbon-14 left as to be unmeasurable.

BACK TO SEA URCHINS

So how could bomb testing validate Ebert's estimates of red sea urchin age? By shaving away thin slices of an urchin's jaw from the most recently grown end back to its root, Ebert thought that, if the urchin had been alive prior to when the bomb pulse occurred, the older jaw slices should show an abrupt drop in the carbon-14 level as it was formed before the bomb pulse. If no such drop was seen, then the urchin must have been born after the pulse. Ebert collected his urchins around 1990, a year after he had originally marked them, so if his age estimations were accurate, he should be able to predict from their size which ones were older than about thirty years. And in fact, just as his tetracycline marking technique suggested, this is exactly what he found.

Let's consider for a moment how impressed we should be by this possibly two-hundred-year-old red sea urchin and what we might learn about the biology of aging from it. Compared to a thousand-year-old or more tubeworm, a couple of centuries may not seem impressive. However, the urchins were not living a mile and a half deep at near freezing water temperature. The average sea temperature in shallow waters off the coast of Washington where Ebert found his longest-lived urchin population was a comparatively toasty 10°C (50°F). So temperature by itself did not dictate a glacially slow metabolic rate. It is only fair to point out, though, that red sea urchins have an exceptionally low metabolic rate compared with other urchins. Another study found that red sea urchins living in these relatively balmy waters had a similar metabolic rate to an Antarctic species living in nearly freezing seawater. So slow metabolism likely plays a role in its long life, as, no doubt, do protective spines that help keep urchins safe from some predators (although they are dined upon with alacrity by sea otters).

Metabolism aside, the main interesting feature of sea urchins—and it is enormously rare as well as interesting—is that they don't appear to deteriorate as they age. That is, older urchins are no more likely to die than younger ones, nor does their reproductive rate fall off even at the oldest ages. If anything, reproduction increases as larger, older urchins have larger, if not necessarily tastier, gonads than their juniors. We don't have many good ways other than mortality and reproductive rate to assess health in urchins. We could measure how much they spontaneously move around in laboratory tanks, which is something we do in mice, but so far no one has done that. What biologist Andrea Bodnar has done, however, is to amputate a few spines and tube feet from young and old urchins and measure how fast they regrew. She found no slowing of regrowth in older urchins, either. The only possible sign of aging found so far is that older individuals of some species had slightly slower rates of cell replacement in some—but not all—tissues than younger individuals.[2] This may be similar to the way hair growth slows with age in people. If so, that is the only sign of aging we have in urchins so far.

CLAMS, OYSTERS, AND QUAHOGS

One day, out of the blue, I received a telephone call from two marine biologists in Wales. As I recall, our conversation went something like, "Hello, there, Dr. Austad. We are marine biologists who study clams that live a very long time. Would you like to collaborate with us to investigate how they do it?"

I get a fair number of crank calls and emails from people who want to live forever—or already know how to live forever and just want my help in spreading the word. Often they also have their own personal theory of the universe and the origin of life they would like to share as well. I try as hard as I can to be polite.

Being polite in this case, I recall my answer was something like, "Possibly. What do you mean by 'a very long time'?"

"Centuries."

I held the telephone away from my ear. It was a transatlantic call. Maybe I had misheard. "Sorry, I thought you said 'centuries.'"

"Yes, that's right—centuries."

I had no idea. A few months later, two researchers from Bangor University—Christopher Richardson, an internationally known oceanographer whose main interest was what long-lived clams could reveal about ancient climates, and Iain Ridgway, a young postdoctoral researcher interested in learning how clams lived so long—sat in my office, explaining in detail.

Actually, I'm sure they didn't say "clams" in our original conversation. They probably said "bivalves," a more general zoological term. Bivalves are mollusks with two shells (valves), joined by a hinge, surrounding their body. Clams, oysters, scallops, and mussels are all bivalves. They are all tasty, too. That was pretty much the extent of my knowledge about them at the time.

Mollusks include many more species than just bivalves. They are the most diverse group of animals in the sea and have been for eons. They are probably the best-known fossil group because their shells preserve so readily. Snails are another large group of shelled mollusks. Most snails are seagoing, but most of us are more familiar with those we've seen around the garden or on a dinner plate with butter and garlic. Octopus and squid are mollusks, too. Even though bivalves have nothing you could reasonably call a brain, octopuses have a very well-developed one and are arguably the cleverest invertebrate. Like orangutans, they can solve puzzles and have been known to regularly escape from their enclosures at night, perform some mischief, and return before morning without anyone being the wiser for quite some time. Octopuses are not long-lived, though. They grow, reproduce once, then die—a life history that salmon made famous.

My first question to the clam experts sitting in my office was, "How do you know how long they live?" I was prepared to hear something about radiocarbon dating and "we know they live somewhere between this long and that long." Instead, I learned that they could tell me the precise age—to the year—of any individual clam. They called it *sclerochronology*—literally, dating things by hard parts. Technically, radiocarbon dating of hard parts is sclerochronology, too, but this is different and more precise. The word was modeled on *dendrochronology*, the dating of trees by their annual growth rings. It turns out that bivalve species residing in waters with seasonal changes in temperature and/or food availability develop annual growth rings in the shell (figure 11.1). Richardson could scarcely conceal his glee in telling me about it. "The bivalve shell is an event recorder," he said, "which gives us a glimpse of sea conditions through the centuries." Because seasonal growth varies from year to year, the rings are sometimes wider and sometimes narrower, resembling a barcode. That barcode allows you to travel back in time via a series of clam shells.

It works like this. Suppose a living clam dredged up from the ocean floor turns out to be a hundred years old. The dredge also hoists up some empty shells that sometime in the past contained clams that—counting their growth rings—also lived a hundred years. But when did the clam in one of

Figure 11.1

External growth lines of a four-year-old clam shell (*Chamelea gallina*). Annual growth rings are marked with arrows. The oldest part of the shell is at the tip, or umbo; the youngest is at the margin. Note the year-to-year variations in ring width, which allow overlapping of dates from shells of different ages. To count annual growth rings most accurately for long-lived bivalves, the shell should be sawn in half, and the internal lines counted. Scale bar = 1 centimeter.

Source: Photo courtesy of Miguel B. Gaspar.

those empty shells die? It might have been last year or several centuries ago. However, if its life overlapped with the living hundred-year-old, it may be possible to align their series of wider and narrower rings and so backdate their birth as well as death dates. If they died, say, fifty years ago, then you have fifty years of overlap with the living clam, and now you have a record of shell growth that extends back 150 years, longer than either individual clam lived. Another empty shell may have rings that overlap with the second one. Now you can go back in time even further. By these overlapping series of rings from successively older shells, Richardson and his colleagues had been able to identify shells of clams that had been born as far back as 649 AD.[3] I told you it was precise. By chemically analyzing individual rings, information about more than a millennium of sea temperatures has been reconstructed!

I say "living clams" a bit ruefully. Bivalve sclerochronology unfortunately requires examining a shell in cross-section. This is a highly technical business that involves sawing through the shell, polishing the sawn surface, etching it, capturing a mold of its etched rings on an acetate peel, and examining the acetate under a microscope. This is possible, of course, only after the inhabitant has been removed, or as we biologists like to put it euphemistically, sacrificed. Because you can't determine the age of a clam until it is sacrificed, we will never be able to analyze the tissues, of Ming, the most famous clam ever.

Ming was an ocean quahog (*Arctica islandica*), unaccountably pronounced *koh'- hog*, also known as the mahogany quahog. Let's just call them *Arctica*, as having studied them for more than a decade now, I feel that we can be on a first-name basis. *Arctica* lives on the continental shelves on both sides of the north Atlantic Ocean from Newfoundland to Cape Hatteras in the west and from Iceland to coastal Spain and the Baltic Sea in the east. It appears to prefer cool to cold water (but not icy water), and it never occurs where the sea temperature is too warm, meaning not exceeding 16°C (61°F). This particular celebrity clam was anointed "Ming"—Ming, the Mollusk—in the press after its great age was discovered because its birth would have occurred during China's Ming dynasty. The alliterative name was a nice touch, too. Ming was born in the year 1499. Putting Ming's age in perspective, at the time of his birth, Leonardo da Vinci had just finished painting *The Last Supper*. Christopher Columbus was in the midst of his third voyage to a continent that he still mistook for Asia. Copernicus had not yet published his radical theory that the earth orbited the sun rather than vice versa. Shakespeare would not be born for another sixty-five years, and the invention of bottled beer lay nearly eighty years in the future.

Ming spent the first part of its youth, as do all clams, aimlessly wandering in the water column, before finally settling down in about eighty meters (260 feet) of water just off the north coast of Iceland, where he survived the remainder of the Little Ice Age and continued on through Iceland's centuries-long transformation from a famine-prone country of a few thousand people to one of the most highly technological countries in the world. Ming also lived through the rise of science until it—science—killed him in 2006 to extract whatever historical information he could provide. His remains were buried at sea. In case you are not keeping count, that made Ming 507 years old and still going strong when he was sacrificed in the service of science.

To clarify a couple of points. I say "he," but we don't actually know whether Ming was a he or she. Also, because the *Arctica* that live around Iceland are smaller than the ones around the United Kingdom, where the particular research cruise that found Ming originated, his great age was not suspected from his size alone. As a result, he (or she) was shucked, and his or her innards unceremoniously chucked overboard. His great age was discovered only later, when his internal shell rings were counted.

Before we could continue discussing Ming and other exceptionally old clams in detail, Richardson informed me that I needed a proper introduction to clam anatomy, so we headed to the supermarket. There we bought half a dozen fresh farm-raised clams and half a dozen wild-caught ones. There were several varieties to choose from—small littlenecks, slightly larger cherrystones, and the largest ones (likely destined for somebody's chowder). The first thing I learned from the clam experts was that these different clams were all the same species, the table clam or *common* quahog, *Mercenaria mercenaria*. They were just harvested at different sizes/ages. The second thing I learned is that fresh clams sitting on a bed of ice in your market are usually still alive. Taking some back to the lab, we plopped them in a seawater aquarium, where they settled to the bottom, cracked open their valves for some oxygen, and waited to be fed.

I had always assumed that clams' bodies were little more than amorphous lumps of flesh. That is what they seem like on your tongue and in your chowder. In fact, they have a complex anatomy, including several large muscles—including the foot (which can be extended from the shell to help the animal move, dig, or push) and the adductor muscle (which closes the shell). The mantle surrounds the entire body, coating the inside of the shell, which it secretes. There are two siphons, one for sucking water in and the other for spitting it back out after removing food particles and oxygen. They even have cells analogous to our white blood cells, called *hemocytes*, circulating in their body fluid, providing immune protection and helping to repair wounds, among other things. These hemocytes can sometimes proliferate out of control, causing the bivalve equivalent of leukemia. Leukemia is only one of several types of cancers that have been reported in clam species. In fact, one type of bivalve cancer has proven to be contagious.[4] How scary is that? You could catch cancer from others like you might catch COVID-19. This is only the third instance of contagious cancer known (the other two are in dogs and Tasmanian devils) and the only one that seems to be able to hop between species and between continents. We may one day learn a great deal about cancer and its prevention from the study of bivalves.

Finally, there is the heart. Who knew that clams had a complex, three-chambered beating heart? It is not exactly the machine-gun-beating hummingbird heart. For long-lived *Arctica*, the heart beats about seven times per minute when relaxing in 15°C (50°F) water but can accelerate to twelve beats a minute in a panic or slow to two beats per minute when doing whatever is the *Arctica* equivalent of meditating. I thought of Ming's heart, which had been beating from sixty-five years before Shakespeare was born until it was scooped from its shell and flung overboard more than five centuries later. How much longer could it have beat?

Nonclam experts always seem surprised by Ming's size. Most people imagine long-lived clams must be enormous, like one of those giant tropical clams the size of a coffee table. Those clams, because they live in warm, shallow tropical water without seasonal temperature variation, do not have growth rings in their shell. However, because they are raised commercially, we know that they are not particularly long-lived—likely fifty to sixty years at most. They simply grow fast in those warm, shallow seas, aided by the energy they get from photosynthetic algae that live in their tissues. Ming, on the other hand, could sit nicely on your palm.

Ming is not the only long-lived bivalve. As a group, they may be the longest-living animals. More than a dozen species have been documented to live a century or longer, including the Pacific geoduck clam (168 years) with its ridiculously long siphon, the freshwater pearl mussel (190 years), and the recently discovered giant deep-sea oyster (*Neopycnodonte zibrowii*), which does not have growth rings but has been radiocarbon dated with all the uncertainly that implies to over five hundred years.[5] In this latter case, *giant* and *deep sea* are relative terms. The largest of these "giant" clams had a shell about thirty centimeters (one foot) long. "Deep sea" in this case means four hundred to five hundred meters (1,300 to 1,650 feet). That's deep enough but nowhere near tubeworm deep. Nor do they live in tubeworm cold. Average sea temperature at the depth where they were found was a balmy 12°C (54°F).

You may be—I hope you are by now—asking yourself why bivalves live so long? Let's run through the list. Ectothermic? Check. Low metabolic rate? Most bivalves hunker down and seldom move as adults, so that's a check, too. *Arctica* in addition has a particularly slow metabolic rate even among bivalves and is one of the slowest-growing species known. It also seems exceptional among bivalves in its ability to survive low oxygen levels, even lower than mole-rats. They have survived two months in the complete absence of oxygen, at least in part due to their ability to reduce their metabolism to a tiny fraction (perhaps as little as 1 percent) of its normal rate for up to a week. Does this ability to survive low oxygen play a role in their exceptional longevity as it may in mole-rats? We aren't sure yet, but I'm betting on it.

Lives in the cold? Well, kind of—half a check. Ming lived in water that was around 6°C to 7°C (43°F to 45°F), which is cold but not deep-sea cold. However, in warmer water, it does seem shorter-lived. In the Baltic, for instance, they don't appear to live longer than fifty years, although it is unclear whether this is due to the Baltic's warmer temperatures (12°C to 14°C, 54°F to 58°F), the low and variable salinity due to all the European rivers that empty into it, pollution of other sorts from those rivers, or its shallowness (average depth fifty-five meters or 155 feet), with all the environmental instability that entails. As noted above, the five-hundred-year-old giant deep-sea oyster lives in at 12°C (54°F). However, we can't discount the possibility that extreme cold may help. There may be clams around cold seeps or on the Antarctic sea bottom that make Ming look like a whippersnapper. So far, we just don't know.

Living in a safe environment? Ah, now we're talking. The ocean, once you get beyond the surface layers, is a relatively stable place to live. The deeper you go, the more stable it is. By stable, I mean safe from abrupt environmental changes. Ocean temperatures on a millennium scale may vary by a few degrees, but compared to climatic variation on land or at the sea surface, this is trivial. Inside a bivalve shell is also a nice, safe place to live. The longer a bivalve lives, the larger and thicker its shell grows, and so the number of potential predators that can break through it is gradually reduced. A third thing that makes things even safer is that many species of bivalves live partially or fully submerged in the muck of the sea bottom. Their siphons can be extended like snorkels out into the open water to feed and resupply their oxygen needs. Feeding and breathing—that is, filtering food particles and oxygen molecules out of water—are pretty much the same activity in bivalves. Burying themselves in the sea bottom may be one factor that helped bivalves survive the Permian Great Dying, that cataclysm of 250 million years ago that killed more than 95 percent of all animal species on the planet. It killed at least that percentage of mollusks as well, except for those bivalves hunkered down on the sea floor. Almost half of bivalve species survived that event.

If these factors help explain the exceptional longevity of some bivalve species, we might expect their opposite to predict short life—even in bivalves. For instance, if there were bivalve species that lived in warm, shallow, less stable surface waters and exposed themselves to dangers by, say, actively swimming (which would also require a higher metabolic rate), you might expect they would be short-lived. Yes, such clams exist. The bay scallop (*Argopecten irradians*) lives in warm, shallow water and actually swims by flapping its valves open and shut like castanets. Bay scallops live only a year or two. That makes their adductor muscle, the part we eat, no less tasty, however.

The nice thing about most of these long-lived—as well as short-lived—bivalves is that they can be brought into the laboratory and studied. My colleagues and I have been doing this for some time. Although we can't yet claim to have discovered the secret of their five-hundred-year lifespan, we have discovered two things worth mentioning here.

Number one is that longevity is related to the ability to resist damage from oxygen radicals. We discovered this during a summer training course in the molecular biology of aging, which I codirected for a decade with geneticist Gary Ruvkun from Harvard. That course was held at the famous Marine Biological Laboratory in Woods Hole, Massachusetts, on Cape Cod. MBL, as it is fondly called, opened in 1888 and has been a summer mecca for biologists ever since. Some are there to teach, some to do research, and some to relax in their elegant summer homes. On any given day in the MBL lunchroom, it is quite possible to bump into any number of past, current, or future Nobel Prize winners. In fact, MBL boasts that as of 2018, fifty-eight Nobel Prize winners have been associated with the lab as students, teaching faculty, or researchers. One year, there were four future Nobel Prize winners teaching in a single summer course.

Because MBL has exquisite seawater facilities for maintaining bivalves and because at the docks where the fishing boats delivered their catch each day we could find any number of bivalve species, I decided one summer to make use of the course's students to investigate how well different bivalve species could survive oxygen-radical stress. The fishing boats sold us some *Arctica*, which we knew could live up to five hundred years. We also bought some table clams, having established by now that they could live up to a century. Then we got several other species that lived about twenty years and our bay scallops, which lived only a year or two. Then we added some oxygen-radical-generating chemicals to their tanks and recorded what happened. The results were remarkable. The normally short-lived scallops all died within two days. The twenty-year-old clams died by day five. It took eleven days for just half of the hundred-year table clams to die, and *Arctica* seemed unfazed. After two weeks, they remained alive and seemed, well, happy as a clam. We tried several other chemicals that damaged cells in various ways with similar results. These observations confirmed what had been seen in standard laboratory animals. Animals that lived longer could tolerate greater assaults from the damaging by-products of life, such as oxygen radicals.[6] Understanding the nature of that greater tolerance could potentially teach us something about how to live longer in good health.

We discovered one other thing about *Arctica* that was delicious in its irony. Even though clams lack anything that could reasonably called a brain, *Arctica* just might hold a therapeutic key to Alzheimer's disease.

One of the main challenges faced by long life is maintenance of precisely folded proteins inside their cells. Recall that for proteins to perform their proper functions (and they provide nearly all the functions inside a cell) requires complex and precise origami-like folding. Over time, proteins lose this precise folding. When they become misfolded, not only can they no longer perform their normal cellular function, but they become sticky and clump together. The familiar plaques and tangles, hallmarks of Alzheimer's disease, are clumps of sticky, misfolded proteins.

Knowing this, my colleague, protein biochemist Asish Chaudhuri, along with graduate student Stephen Treaster and I turned our eyes seeing how well *Arctica* could resist misfolding of its proteins.

We applied a number of well-known methods for purposely misfolding proteins to the liquid cellular extract of clam species that lived only up to seven years, along with another species that lived up to thirty years, another living a hundred years, all the way to *Arctica*, the five-hundred-year species. We found that no matter how many ways we tried, proteins from the *Arctica* resisted our attempts to misfold them. One reason for this could be because *Arctica*'s proteins were themselves intrinsically resistant to misfolding. Another, more interesting reason was that there may be some molecules in *Arctica*'s extensive protein protection machinery that made it superior to such machinery of other species. If that were true, then proteins from any species, even humans, might be made more resistant to misfolding. We might be able to use it to prevent protein-misfolding diseases like Alzheimer's disease. This latter idea turned out to be true. *Arctica*'s protein protection machinery worked better than any other clam. In fact, it worked better than a similar extract from human tissue on any protein we tried.[7] We even tried it on human A-beta, the protein that makes Alzheimer's plaques.

At this point, we began to get very excited. If we could isolate the molecule or molecules in *Arctica's* protein maintenance machinery that was responsible for this exquisite ability to resist misfolding, we might be able to use that knowledge to develop treatments for protein-misfolding diseases like Alzheimer's and Parkinson's disease.

Now that you're poised for the big reveal, I have to report that for the past seven years, we have been searching for the secret of *Arctica's* ability to prevent protein misfolding. We have been highly successful at discovering a number of things that it isn't. Alas, at least to date, we have not been able to figure out what it is. We will keep working on it, though. Science is seldom quick or easy.

12 FISHES AND SHARKS

Why are there so few fish in the sea? I suspect that is a question few readers have asked themselves, mainly because we generally assume that the number of fish in the sea is almost limitless. This is evident from the way we overfish species after species after species. But what I mean by this question is this: why does the sea, which contains 97 percent of all the water on the planet, contain slightly less than half of all fish species?[1] That fact is even more puzzling than it may seem at first glance because most—more than 99 percent—of fresh water is tied up in glaciers and snowfields. So the available fresh water in rivers, lakes, and streams where fish can live is much, much less than 1 percent of the water in the ocean, yet there are slightly more fresh-water fish species than there are marine species. What gives?

The answer is relevant to understanding the longevity of fish species, and please understand that I'm using *fish* here in a broad, traditional sense that includes sharks and rays. The ocean is stable, almost unchanging, compared to lakes, rivers, and streams, which come and go as the climate changes and tectonic plates shift across geological time. The ocean is several billion years old. Most lakes have existed for no more than a few thousand years; the oldest ones, for a few million years. Rivers and streams are typically even younger. Streams and lakes will shrivel or disappear during millennial droughts. They swell, deepen, and merge to form new lakes during periods of greater rain. Ten million years ago, the Sahara Desert was a swamp. Twenty thousand years ago, America's Great Lakes were buried under a mile of ice, as is much of Antarctica today. Even the oldest and deepest lake in the world,

Siberia's Lake Baikal, is a mere 25 million years old, a whippersnapper compared to the ocean.

Stability is the enemy of new species formation, and change is its friend. There are slightly fewer than fifteen thousand species of fish in the ocean, most of them living in coastal waters because, as we've seen, the open ocean is a nutritional desert. Slightly *more* than fifteen thousand fish species live in the world's lakes, rivers, and streams. A single lake, Africa's Lake Victoria, before a recent massive wave of human-caused extinction, contained more than five hundred fish species. Lake Victoria is a mere 400,000 years old, but it has been anything but stable during that time. It has shrunk and fragmented into many small lakes and ponds at times and grown again when the rains returned and the smaller lakes and ponds re-fused. Lake Baikal, with ten times the amount of water and more than fifty times the age of Victoria, is much more geologically stable but has only sixty-five species of fish.

Species form when populations become fragmented. Isolated populations inevitably drift apart in idiosyncratic ways over time, and when they eventually connect again, they may differ enough so that they can no longer interbreed. The once isolated populations have become new species. Islands are crucibles of species formation because, by definition, their populations are isolated. Island archipelagos foster swarms of new species. Considered from the perspective of fishes, the individual puddles and ponds of Lake Victoria during millennia of drought were habitat island archipelagos, and when the rains returned, newly evolved species met one another for the first time.

Stability may be the enemy of new species formation, but as we now know, it is the friend of slow aging. So it should not be surprising that although it contains fewer fish species than fresh water does, the ocean is home to our longest-lived fishes.

Not that there are not some reasonably long-lived freshwater fish.

A lot of effort has gone into developing methods for estimating the age of wild fish. Knowing how long fish live is critical for developing accurate population growth models. Accurate population models are necessary for implementing sustainable fisheries strategy. At the turn of the twenty-first century, fish represented a sixth of the animal protein consumed by humans, and that fraction has since grown rapidly, even as the global population has increased by more than a 1½ billion people in the past two decades. Global food security will require sustainable fisheries practices.

Fish, like sea urchins and bivalve mollusks, continue to grow as they age. So the biggest fish, "the one that got away," is often the oldest. This is true *within* a fish species but not necessarily *between* species. Although the general trend is for larger species to be longer-lived, we have seen that is a general pattern and not a hard and fast rule. Giant tropical clam species, for instance, live a fraction as long as the much smaller ocean quahog. A seven-gram (quarter-ounce) bat can outlive any dog.

Species with *indeterminate* growth, as it is called, invariably grow progressively slower as they become older and larger. So a very large individual will clearly be older than a small individual, but there is lots of variability in growth rate among individuals. Determining which of several very large individuals is the oldest is harder to determine—which brings us to fish scales. Fish scales and other hard parts of their bodies also continue to grow as the fish grows. When growth varies seasonally, annual or biannual growth rings are often visible, just as they are in clam shells. Counting growth rings in scales is the oldest, easiest, and least invasive way to determine a fish's age. However, all growth rings are not annual rings (figure 12.1). Maybe they follow nutrient pulses on lunar cycles or daily cycles or have other structural purposes and appear at unpredictable intervals. To use growth rings to determine age, you need some individuals with well-documented ages to calibrate the ring counts. Not having such a standard can lead to wildly inaccurate ages. Even when there are annual rings, at great ages, growth may be so slow that individual rings become so tightly packed they are difficult to distinguish. Because of this complication, Ming the Mollusk was originally determined to be a paltry four hundred years old. Determining his or her real age required an improvement in growth-ring detection technology as well as comparison with overlapping rings in younger individuals that died centuries ago.

The freshwater fish with the best-known reputation for exceptional longevity is, no doubt, the carp. In Aldous Huxley's 1939 satirical novel *After Many a Summer*, about youth obsession among the Hollywood movie crowd, it had been discovered that eating raw carp entrails could stave off aging. Carp

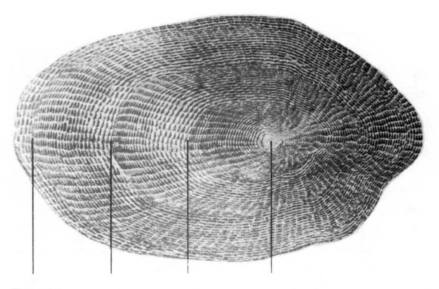

Figure 12.1

Growth rings in a fish scale. It is possible to determine fish age from growth rings in fish scales if fish of known age are available to calibrate the rings' meaning. Lacking calibration, age can be dramatically mischaracterized. For instance, the fish scale above has about ninety rings. Assuming they were all annual rings would dramatically overestimate the age of the fish, if we didn't know its true age. The lines point at the annual rings. This scale is from a four-year-old, forty-five-centimeter (eighteen-inch) haddock.

Source: F. E. Lux, "Age Determination in Fish," US Fish & Wildlife Service. Fishery Leaflet No. 488, US Fish and Wildlife Service, 1959.

and humans who ate their raw entrails could live up to two hundred years. There was one unfortunate side effect, though. Over the course of their exceptionally long lives, people who ate carp entrails eventually were transformed into creatures resembling fetal apes. For some people I've met, that might be an acceptable price to pay for longer life. We were, after all, fetal apes once ourselves.

I say "carp" as if that were a meaningful name. In fact, it is a zoologically vague word that could refer to dozens of species of large freshwater fishes. Zebrafish and minnows could both be called carp as they are in the same fish family. However, it is the beautiful koi carp, domesticated version of the common carp (*Cyprinus carpio*), that gets—and deserves—the attention.

Beginning in China more than a thousand years ago, various species of carp were domesticated—that is, purposely bred in artificial ponds for food and ornamental purposes. The goldfish is one product of that ornamental breeding. In the nineteenth century, carp in Japan began to be selectively bred for various colors and color patterns. Today, they are bred by hobbyists worldwide and come in red, blue, cream, yellow, black, white, and mixed-color varieties. In 2018, a particularly beautiful specimen was reputedly sold for $2 million.

Koi can grow up to about a meter (thirty-nine inches) in length and fifteen kilograms (thirty-five pounds) in weight. They prefer relatively warm water (15°C to 25°C, 59°F to 77°F), which suggests that we might not expect them be particularly long-lived. In fact, they are well-documented to regularly live twenty to thirty years, with a few occasionally approaching as long as fifty years, which is not bad for a warm-water, freshwater fish.

So given that many koi breeders over the years have determined that twenty to thirty years is a pretty typical koi life, what do we make of the tale of a scarlet koi named Hanako that lived to the reputed age of 226 years before dying of an unknown ailment in 1977?

Let's consider the evidence. The first piece is that Hanako's last owner, Komei Koshihara (and, of course, an unknown number of previous owners, most recently members of Koshihara's family), said so. In 1966, he claimed on a national radio broadcast in Japan that Hanako was hatched in 1751. Hanako was a compact seventy centimeters (twenty-eight inches) in length and weighed 7.5 kilograms (16.5 pounds) in 1966, not as big as you might expect a 215-year-old (at the time) female koi to be. The second piece of evidence was the growth ring count of two of Hanako's scales by Masayoshi Hiro of the Laboratory of Animal Science at Nagoya Women's College. Koshihara, the fish's owner, I might note, was president of that college—Hiro's boss, in other words. In the koi hobbyist world, Koshihara and Hanako became famous overnight.

Now let's pause briefly. In science, it is generally accepted that extraordinary claims require extraordinary evidence. Checking back on the transcript of the original radio broadcast, it seems that it took two months for Hiro to determine Hanako's age from her scales. For someone experienced at let's call it "scale-ology," this would seem like a long time to do something as seemingly simple as counting fish scale rings for your boss. In fact, we have no evidence that Hiro had any expertise in fish age determination, and as you can see from the rings in figure 12.1, there is plenty of opportunity to inflate age based on rings if you have no known age comparison. Also, when asked to determine the age of four other carp in Koshihara's pond, he came up with two that were 139 years old and three others at 149, 153, and 169 years. Considering that no other koi older than about fifty years has ever been reported before or since, this either says a great deal about the healthy environment of Koshihara's pond or Hiro did not know what he was looking at. A third possibility, which I hesitate to mention, is that Hiro may have been eager to please and make famous in koi social circles his boss, who had brought him these scales precisely to verify the exceptional age of his fish. Put another way, if the average age of koi is around twenty-five years, as numerous sources suggest, what are the chances of finding an individual living nine times the average and a number of others five to seven times longer-lived than average in a single pond? It would be like stumbling on a seven-hundred-year-old human in your local social club, with five of her neighbors still alive at four hundred to six hundred years.

As I say, the longest-lived fish all live in the ocean.

But some ocean-going fish live in freshwater, which brings us to the longest-lived quasi-freshwater fish—the sturgeons.

STURGEONS

I should probably qualify and clarify the statement that sturgeon are freshwater fish. They are fish capable of living in fresh water. They *must* lay their eggs in fresh water. However, if given a chance, they will spend most of their lives prowling the brackish water of coastal bays and estuaries where river and sea water mix. They do this along the east and west coasts of North America and many of the coasts and inland seas of Eurasia. However, to spawn, like salmon, they must travel up rivers to find fresh, clear water with shallow rock or gravel bottoms, where they lay eggs or broadcast sperm (depending on their sex) before returning to the ocean. Over the eons, as sea levels rose and fell and tectonic plates shifted, some sturgeon became trapped, land-locked in large inland lakes. As they already had the ability to survive in fresh water, there they remained, needing only to swim from the lakes into the clear running streams and rivers that fed them when it came time to spawn. The world's largest lake by surface area, the slightly salty Caspian Sea, contains six of the approximately twenty-five species of the world's sturgeons.

Sturgeon range in size from the twenty-seven-centimeter (ten-inch), fifty-gram (two-ounce) dwarf sturgeon (*Pseudoscaphirhynchus hermanni*) of Central Asia to the 3½-meter (ten-foot), one-metric-ton (2,200-pound) Beluga sturgeon (*Huso huso*), the world's largest "freshwater" fish. These days, Beluga sturgeon are found mostly in the Caspian and Black seas. Beluga sturgeon roe—eggs, that is—when salt-cured is sold as Beluga caviar, which will set you back about $1,000 an ounce at today's prices. As up to an eighth of the body weight of a gravid female Beluga sturgeon consists of roe, this is without question the most commercially valuable fish in the world.

Incidentally, Beluga sturgeon should not be confused with Beluga whales, which produce no caviar, live only in the ocean, and do not live nearly as long as Beluga sturgeon. Beluga is the anglicized form of the Russian word for *white*, the dominant color of both species.

Beluga is not the only sturgeon from which caviar is made. In fact, in the early twentieth century, before it became a delicacy that only the wealthy could afford, American caviar dominated the markets of Europe. It was so plentiful and cheap in those days, being made from the roe of lake, short-nose, and Atlantic sturgeons, that it was given away to bar patrons to stimulate thirst, somewhat like peanuts are today. However, all species of sturgeons have been threatened by overfishing, dams that block access to spawning grounds, and water polluted by industrial or agricultural runoff. It is one of the most heavily regulated fisheries in the world but is still in a delicate state. Most caviar is now produced from sturgeon born in hatcheries and released into the wild.

Also, largely because of the caviar industry, we know a great deal about how long sturgeons live and the tragic opportunity we may have lost for understanding that great longevity. As with mammals, species size counts.

Dwarf sturgeon can live up to an easily determined six years. Beluga and other large sturgeons live a long time—a very long time.

Sturgeons, alas, do not have scales, so "scale-ology" cannot be used to determine age. However, they do have hard parts that grow along with them, and like scales, those hard parts have growth lines. For nearly a hundred years, sturgeon have been aged by examining growth lines in cross-sections of their pectoral fin rays. Fins rays are those bony spines that stiffen fish fins. This is a reasonably accurate method of determining a sturgeon's age, having been calibrated with both known-age fish and by the bomb pulse. If anything, such counts will *underestimate* age in older fish where the growth lines are so close together they seem to merge.[2]

To date, the oldest sturgeon of any species aged by fin rays is a 152-year-old lake sturgeon (*Acipenser fulvescens*).[3] Caught in 1953 in Lake of the Woods, an island-dotted lake spanning parts of Minnesota, Ontario, and Manitoba, this was a good-size but not giant fish of this species. She was a bit over two meters (eighty-one inches) long and weighed ninety-eight kilograms (215 pounds). Before that 152-year age swims out of view, let me point out that this fish would have been born in the same year that Thomas Jefferson defeated John Adams and Aaron Burr to become the third American president. So making this very clear, at least one individual lake sturgeon lived from the presidency of Thomas Jefferson until it was murdered (perhaps we can call it justifiable homicide) by a fisherman during the presidency of Dwight Eisenhower. Also, putting its longevity in mammalian perspective, that 152 years *in the wild* would represent a longevity quotient of a very bat-like 8.5.

Sturgeons, like many politicians, are bottom feeders, scooping and sucking snails, leeches, insect larvae, mussels, and even small fish from the sediment. Lake of the Woods is only about sixty-five meters (210 feet) at its deepest, so while it is ice-covered in the winter, its surface waters can warm to 20°C (68°F) in the summer. Lake sturgeon spawn in rivers where spring water temperatures range from 13°C to 18°C (55°F to 64°F). So exceptionally low metabolic rate due to exceptionally cold water does not seem to play a role in the exceptional longevity of this species.

Lake sturgeons, as you might predict, proceed slowly through all stages of life. They grow slowly, spawning for the first time at around ages fifteen and twenty-five years for males and females, respectively. They reproduce slowly. Females spawn every four to nine years. And of course, they age slowly. Interestingly, although the 152-year-old lake sturgeon was big, it was nowhere near the size and possibly nowhere near the age that some Beluga sturgeons reach.

Beluga sturgeons also proceed slowly through life. Males spawn for the first time at around age fifteen years, and females at around twenty years. Today, commercially harvested Belugas are small by historical standards, on the order of 1½ to three meters (five to ten feet) long, weighing between twenty to 250 kilos (forty-four to 550 pounds). Notice these sizes are smaller than what I described for this species earlier. That previous description was Beluga size averaged over the past fifty years or so. Because of selective fishing for the biggest and therefore oldest sturgeons, Beluga elders have become rarer and rarer. An average adult Beluga today is only about thirty-five years old, and the oldest ones are likely only fifty to fifty-five years old. However, as Russian fisherman can tell you, earlier in the twentieth century, century-old Belugas were common[4] (figure 12.2). The oldest, in fact, was documented by fin ray analysis to be 118 years old, which makes me wonder about the age of the biggest Beluga, caught nearly two centuries ago. That female was famous in her day. She was landed in 1827, well before fin-ray analysis and also well before regulated fishing. She measured 7.2 meters (twenty-four feet) in length and weighed 1,571 kilograms (3,463 pounds). The precision of these measures from two centuries ago indicates how valuable Beluga sturgeons were even then. A little back-of-the-envelope calculation reveals that such a Beluga would contain about seven thousand ounces of roe, which turned into high-quality Beluga caviar, would fetch about $7 million at today's prices. Alas, we will never know how long that Beluga had to live to reach such a massive size.

Why are sturgeon—large sturgeons, anyway—so long-lived? They don't live in the most stable of environments. However, they are protected from external dangers, at least in the form of predators. Like the giant

Figure 12.2
Three Beluga sturgeons caught in the Volga River in 1924. Notice the size range. Because of selective fishing for the largest sturgeons, individuals the size and age of the large one here no longer exist.

tortoises, their size and armor protect them. Armor for sturgeons comes in the form of pointed bony scutes, looking somewhat like small teeth, that line their sides and back. And of course, they are ectothermic with a slow metabolic rate, shown by their slow growth. But metabolism is not likely the whole story. Sturgeons spend a reasonable amount of their time in shallow, warmish water, where their metabolism would speed up. There must be more to the sturgeon longevity story. Also, as far as we know, they don't age. The largest, oldest females are still producing roe. The bigger they get, the more they produce. Males at the oldest ages recorded are still producing sperm. It is unfortunate that we are unlikely to ever see the very oldest sturgeons again.

ROCKFISHES

When I said that the very oldest fish live in the ocean, I was thinking of the rougheye rockfish (*Sebastes aleutianus*), the oldest fish known (if we ignore sharks for now). Rockfishes, as their name implies, live among rocks on the sea bottom. Some species, such as the rougheye, live at depths up to nearly nine hundred meters (three thousand feet), and others I have swum alongside at a depth of only ten meters (thirty-three feet) or less when scuba diving in Puget Sound. We may have a considerable amount to learn about aging, possibly even discovering some tricks about extending human health, from rockfishes as a group.

I say this because this group of about a hundred closely related species, all in the genus *Sebastes*, has an amazing range of longevities from little more than a decade to over two hundred years. Understanding the biological underpinning of short- versus long-lived species is a marvelous tool for understanding the biology of aging itself. At least six rockfish species live a century or more, but there are also plenty of short-lived species. By now, I hope the first question in your mind is, How do we know how long rockfish live? The answer is that the most accurate technique of aging bony fish turns out to be counting growth lines not in scales or fin rays but in otoliths. Otoliths are pebble-like structures floating in the fish's inner ear that help it sense gravity and movement. Humans have otoliths as well—two in each ear. Most fish have three. You never notice how important sensing gravity is until you lose it. I once developed an inner-ear problem that made everything spin crazily—the room, the ceiling, everything, whether I was standing or lying in bed. Fortunately, after several weeks of drugging myself to combat nausea from motion sickness, it went away. That experience gave me a lasting appreciation of my otoliths, though.

In fish, otoliths are thought to help keep them upright and to be involved in hearing. Otoliths grow as fish themselves grow and seem to be a bit more reliable than scale or fin-ray growth lines for estimating fish age. This has been confirmed on known-age fish and by several radiometric methods, including the carbon-14 bomb pulse.[5] Like any of the other sclerochronological techniques, it requires some expertise in preparing the otolith for ring counting, but it is now commonplace in fisheries biology.

Rougheye rockfishes, so-named for the spines below their eyes, can be pink, tan, or brown and sixty to ninety centimeters (two to three feet) long. They would definitely not be one of the rockfishes I saw when scuba diving as they live on the sea bottom far deeper than scuba diving is possible, most are found at 150 and 450 meters (500 to 1,500 feet) but individuals have been caught considerably deeper. Along with at least fifty other rock-fish species, it lives in the northeast Pacific from California to Alaska. The water temperature at the depth where the rougheye lives ranges from 0°C (32°F) to about 5°C (41°F). At 205 years, the rougheye is the longest-lived rockfish. In mammalian terms, that would be a longevity quotient of 14, far greater than any mammal. Doing the past-president comparison once again, it is possible that a rougheye currently lurking among rocks on the sea bottom was born during the presidency of James Madison at the time that the Battle of New Orleans was fought. One could possibly learn something about aging by compare the rougheye with the deliciously named Bocaccio rockfish (*Sebastes paucispinis*), which is roughly the same size but lives only fifty years. Other interesting comparisons might be with the one-third smaller quillback rockfish (*Sebastes maliger*) or even the seven-inch (eighteen-centimeter) Puget Sound rockfish (*Sebastes emphaeus*), both of which I definitely saw scuba diving. These species live ninety-five years and twenty-two years, respectively. Of course, their longevity is reflected in the age when they begin to reproduce. Rougheyes reach reproductive size in about twenty years, the quillback in eleven years, the Bocaccio in eight years, and Puget Sound rockfishes rush to reproduction in one to two years.

Recently the genomes of eighty-eight rockfish species were investigated making just this sort of short- versus long-lived species comparison.[6] One set of genes that showed signs of having been under natural selection in the long-lived species only was those involved in DNA repair. Alert readers will note that DNA repair genes were also identified in the giant tortoise genomes. Hence there are good reasons theoretically and in actuality to think that DNA repair ability is a necessary feature of long life. Unfortunately, and this is why I say genome sequencing is a good and necessary first step toward understanding exceptional longevity but nothing more than a first step: we have no idea whether or not long-lived rockfishes or giant tortoises actually repair their DNA better than humans do. Moreover, if they do, exactly how do they manage to do it? That level of deeper understanding requires deeper biological investigation. Alas, studying the genome of the rougheye, or possibly their cells in a dish, may be the only thing we will be able to study in that species. Living at such great depth under as much as 150 atmospheres of pressure, they are inevitably dead from what is euphemistically called *barotrauma* (exploding swim bladder) by the time they are hauled to the surface.

SHARKS

The fact that a shark is the longest-lived fish should not come as a complete surprise as sharks are some of the biggest fish in the sea. In fact, the whale shark (*Rhincodon typus*) *is* the absolute biggest, with one specimen reported to reach twenty meters (sixty-six feet) in length and weigh thirty-four tons, making it nearly three times as long and more than twenty times as heavy as the largest sturgeon. Whale sharks are filter feeders, gulping massive amounts of water with a mouth that should properly be called a maw, if anything should be, filtering out whatever small critters—squid, krill, or fish—happen to be in that water. There has never been a human thus swallowed, although some have speculated that a whale shark, rather than a whale, may have been what swallowed Jonah in the biblical tale. Whale sharks do live in the Mediterranean, so the geography is sound. However, despite the fact that their gape is plenty large enough to encompass a person, their esophagus is only inches wide. So Jonah would have had a difficult time making it to the stomach, much less living there three days before having to make the return trip through that narrow passage when he was vomited back out. They are spectacular fish even if they don't eat people, and whale shark tourism is an increasingly popular activity.

The size mentioned above is extreme even for a whale shark. An average one runs more in the six- to seven-meter (twenty- to twenty-three-foot) neighborhood. Even more than longevity records, size records—especially size records of fish, especially size records of *charismatic* fish—are prone to exaggeration.[7] If this is the biggest fish in the sea, however, it does make you wonder how old might it be.

Sharks pose some special challenges in determining their age because they lack scales, otoliths, and bony fin rays. In fact, they lack bony anything.

Their skeleton is composed entirely of cartilage. Their fin rays are cartilage fibers, which are the stringy things in shark fin soup, a culinary delicacy that I have not tried but that must taste better than it sounds. Lacking these earlier validated measures of bony fish age, researchers have recently validated one other thing—growth bands in the vertebrae, again by calibrating age relative to the carbon-14 bomb pulse, as a reasonably accurate measure of age in some, if not all, shark species.[8]

Unsupported speculation has whale sharks living as long as eighty, a hundred, and even 150 years. However, vertebral growth band analysis on a small sample of twenty individuals found none to be older than fifty years. This half-centenarian was a ten-meter (thirty-three-foot), seven-thousand-kilogram (15,500-pound) female that died after becoming entangled in fishing gear. This gives her a longevity quotient of 0.9, extraordinarily short-lived for an ectotherm. However, as the largest whale sharks are nearly twice this length, fifty years is likely a serious underestimate of how long they really can live. The best evidence of their greater longevity than fifty years is that whale sharks may not begin reproducing until they reach eight to nine meters in length, so this unfortunate female may have been a youngish adult. Knowing the age of only twenty individuals mean not knowing much about how long a species might live. Imagine the chance of catching an exceptionally old human if we plucked only twenty random people off the street? So how long whale sharks might be able to live is something we just don't know yet. What we can say is that any evidence of exceptional longevity is lacking for this species so far.

The great white shark (*Carcharodon carcharias*), which despite what the movies and your nightmares may say, averages only around four meters (thirteen feet) in length and about a thousand kilograms (2,200 pounds) in weight. It must be said, however, some reasonably credible reports describe individuals almost twice this long and three times this weight. If any fish has charisma or perhaps negative charisma, it is the great white. That name, by the way, may be part of its negative charisma and its unwarranted reputation for being a man-eater. Shark scientists typically refer to them more prosaically as simply white sharks.

A bit about that reputation. I certainly am fascinated by this species. There is unforgettable footage in David Attenborough's *Planet Earth* series of a great white exploding out of the water to grab an unfortunate seal. My wife and I once scheduled a cage dive—where you hover underwater inside a protective cage—off the coast of South Africa to observe them in action. Alas, the dive was aborted due to bad weather. So yes, it is a spectacular predator, fearsome to look at, but let's put their man-eating in perspective. Yes, they do attack humans more frequently than any other shark species, probably due to mistaken identity, but those attacks are still exceedingly rare. There are only about eighty shark attacks on humans per year globally, and few of these are fatal. In 2019, for instance, a down year for shark attacks, there were sixty-four worldwide, only two of them fatal. The United States averages less than one fatal shark attack per year. This means you are much more likely to be killed by lightning, bees, cows, or deer than by a shark.

How long do they live? Again, using vertebral growth lines validated by radiocarbon dating, only a small sample of four great white sharks of each sex have had their ages measured.[9] The oldest was a male, estimated at seventy-three years, a longevity quotient of 1.1. The oldest female was forty years old, LQ of 0.9, eerily similar to the whale shark. Both of these were approximately five meters (16½ feet) long, suggesting that females grow substantially faster than males, at least in the northwest Atlantic Ocean, where these sharks were taken, and also suggesting that the largest great whites, which can reach as much as seven meters (twenty-three feet) in length, could be a lot older. In another study of vertebrae from eighty-one sharks and information on their reproductive status when they were caught, researchers determined that male great whites begin reproducing at around age twenty-six and females at around age thirty-three, which suggests they may live an exceptionally long time.[10] Also, that may seem like a long adolescence, but it is nothing like what we are about to encounter.

The current queen of longevity among fishes is the Greenland shark (*Somniosus microcephalus*). Greenland sharks live in colder water than other sharks. They also have the most northerly distribution of any shark, being found at times cruising below polar ice in the Arctic Ocean. They have been found as far south as the Gulf of Mexico, although I should note for those who swim in the Gulf and already worry about sharks, that their sighting in the Gulf was at more than a mile deep, where the water temperature was just 4°C (39°F). It is also among the largest sharks, coming in just behind the great white in size, averaging about three meters (ten feet) long and 320 kilograms (seven hundred pounds) but with records of animals more than twice that length.

The Greenland shark differs in one key feature related to longevity from whale sharks and great whites. It lives in much colder water. Those other species are both found in relatively warm waters and are more active, meaning a relatively faster metabolism, relative to Greenland sharks. Whale sharks are seldom found in water cooler than 21°C (70°F), and great whites frequent waters between 12°C and 24°C (54°F to 75°F). On top of that, great whites are one of the few fish that are somewhat endothermic. They can warm their internal organs as much as 5°C to 10°C (9°F to 18°F) above that of the surrounding water when necessary.[11] Great whites also migrate great distances.

Greenland sharks are large, pure, ectotherms that live year round in water that is within a few degrees of freezing. Large size and cold water for an ectotherm mean low metabolic rate. Indeed, they even move excruciatingly slow (excruciating unless you happen to be a potential prey item). Researchers monitoring the activity of wild Greenland sharks with sophisticated accelerometers off the coast of Norway found their average swimming speed in 2°C to 3°C (34°F to 36°F) seawater to be about one-third meter (that is, thirteen inches) per second, and the fastest speed attained over a twenty-four-hour period was about two-thirds meter per second.[12] To put this speed in perspective, the average walking speed of healthy people in their eighties, a trait used by geriatricians to assess frailty, is a blazing one meter per second.[13] Yes, a normal eighty-year old walks faster than a Greenland shark swims. See what I mean by excruciatingly slow?

You might wonder how a mega-carnivore catches prey if it moves so slowly. Greenland sharks have been found with a range of prey in their bellies including various fish, squid, other sharks, and seals. Part of the answer may be that they are excellent scavengers. That would explain the polar bear parts and the whole reindeer found in one Greenland shark stomach. The other part of the answer may be that seals sleep in the water. Swimming at a glacial pace may allow them to approach sleeping prey by stealth. Good thing that due to their low metabolic rates, these sharks can survive for as long as a year on as little as a meal of one juvenile seal, a tiny fraction of what a great white requires.

Greenland shark age has been estimated in a unique and necessarily innovative fashion. Due to the constantly cold water in which they swim, they do not have growth rings in their vertebrae. Their age has been estimated using, of all things, the core of their eye lens. The core of the eye lens of sharks and other species forms prenatally. As the shark grows, so does the lens by adding new layers around the core, something like layers of an onion. But the core remains chemically as it was at birth.

The innovative aging technique was to examine shark eye lens cores for evidence of the carbon-14 bomb pulse.[14] Researchers analyzed lenses from twenty-eight female Greenland sharks taken during scientific sampling from 2010 to 2013 and came to a startling conclusion. Two of the sharks appeared to be well over three hundred years old!

Given this extraordinary claim, let's take a magnifying lens to the evidence. The twenty-eight sharks ranged in size from less than a meter to five meters (sixteen feet) in length. One shark—and everything else is based on this one shark—was 2.2 meters (seven feet) in length, medium size in this sample. Chemical analysis of the eye lens of that animal indicated it was born very close to the peak of the bomb pulse in the early 1960s, suggesting that it was about fifty years old when killed. Assuming that 2.2 meters was the size of an approximately fifty-year-old shark, the researchers then made a series of mathematical assumptions about growth rate, size at birth, and carbon-14 levels in the ocean and the eye lens of bigger sharks, to calculate that their two largest sharks were 335 and 392 years old, give or take a century. Yes, the margin of error was that large. This, by the way, is a longevity quotient of 11.7—big, but not rockfish big. If that weren't mind-numbing enough, they also calculated from the size of the smallest female Greenland shark found to be reproducing when captured that reproduction in this species begins at about 156 years of age, again give or take a few decades as the margin of error. Whether you give or take, that is a prolonged adolescence!

As I said before, in science extraordinary claims require extraordinary evidence. Estimating that Greenland sharks live nearly two centuries longer than any other known vertebrate is certainly an extraordinary claim. Given the host of assumptions these age estimates require, the calibration from a single animal that appeared to be born at the peak of the bomb pulse, I would definitely not call what we have at present extraordinary evidence. I would call it *reasonable* evidence, perhaps. The model assumption that individual age and size are deterministically linked is something that we know is not true in other fish species. In fact, the length difference between the two sharks estimated at 335 and 392 years old was only nine centimeters (3½ inches). That seems to me a small difference on which to base a fifty-seven-year age difference estimate, but that may just be my habitual scientific skepticism at play. The most difficult aspect for me to swallow (and this is on the logic of it, not on the evidence) is that nature would allow any animal to wait more than 150 years to start reproducing. But you go with the best available evidence, and so far, this is the best available evidence for this presumably very long-lived species.

Despite these caveats—and several shark researchers have advised me to take these estimates with a large grain of salt—this great longevity does not seem to violate the expected metabolic profile of such a large, sluggish, ectothermic animal living in such a cold environment. So I'm provisionally accepting, at least until better evidence is available, that the Greenland shark is indeed the longest-lived, slowest-maturing, not to mention slowest-swimming vertebrate on the planet.

Before we leave these large shark species, I want to remind readers that one of the inherent dangers in being big, particularly when combined with long life, is increased cancer risk. The more cells in a body, the greater possibility that one will eventually convert to a cancer cell with potentially fatal consequences. The more time for this to happen, the greater the risk. If elephant cells converted as easily as mouse cells, all elephants would be dead of cancer in a year or so. Nature has apparently found solutions to this intrinsic risk, such as the multiple copies of the TP53 tumor-suppressor gene that protect elephant cells. Consider this: whale sharks grow to five times the size of an elephant and so have roughly five times the number of cells. That, by the way, is more than five hundred times the number of cells that we humans have. Even though the oldest whale shark examined to date was only fifty years old, we know this is a considerable underestimate of that species' longevity. Therefore, whale sharks and even Greenland sharks, which have roughly six times the number of cells as a human but live possibly four times as long, both may have a great deal to teach us about how to avoid cancer if we do the research to figure out how they do it.

Why a strange mammal—a wolflike predator with deerlike hooves—would gradually abandon the land where mammals had been wildly successful for more than 15 million years after the massive asteroid impact that finished off the dinosaurs, pterosaurs, and most other species and would return to the water is a mystery we may never fully understand. Perhaps it had to do with an abundance of fish prey—fish recovered quickly from that massive extinction event—potentially offering quick and easy meals with less competition than on the land. Maybe under water was a refuge from terrestrial danger as it is for today's fanged deer of Africa. Whatever those aquatic opportunities may have been, once freed from the bounds of gravity by the buoyancy of water, over the next tens of millions of years the descendants of that strange species underwent extraordinary change, losing their hind legs and modifying their front legs into flippers and their tails into propulsive flukes. Their nostrils migrated to the top of their heads. Their ears were reconfigured to pick up sounds traveling through water rather than air. For some species, that reconfiguring included developing an ability to hear high-pitched sounds used for echolocation, like bats. Although never losing their predilection for carnivory, other species did away with teeth entirely, learning to gulp massive numbers of small prey using a flexible filter called *baleen*. Some species emerged as the largest animals ever to exist.

The evolutionary descendants of that hoofed predator today comprise about ninety species that scientists call *cetaceans*. The rest of us call them variously *dolphins*, *porpoises*, or *whales*. The smallest of them, Heaviside's dolphin (*Cephalorhynchus heavisidii*), is the size of a small person. The largest, the blue whale (*Balaenoptera musculus*), can weigh as much as thirty African elephants, is half again as long as the biggest whale shark ever recorded, and has a heart that has been described as the size of a Volkswagen. Most live in the ocean, although a few are riverine. Some ply coastal waters, and others are found in the open ocean where they can dive more than a mile (sixteen hundred meters) deep in search of prey. None, however, have been freed from the necessity to extract their life-sustaining oxygen from air rather than water or to heat their bodies from within to a body temperature as high as dozens of degrees above the surrounding water. These limitations have significance for their longevity.

There may be no more charismatic group of animals than whales. People may be willing to pay to see a bestiary of exotic animals up close in zoos, but people pay more than $2 billion annually just for the *possibility* of glimpsing whales, emerging from the sea momentarily, from a distance. For several years, I served on the Massachusetts State Right Whale Advisory Committee. Right whales are severely endangered, and I am proud to say that Massachusetts was the first state to establish a rule making it illegal for any vessel to approach a right whale closer than five hundred yards. I recall the passion with which some people love whales. One exceptionally committed civilian member of that committee proposed in all seriousness that violating that five hundred yard rule should be a capital crime.

Whales are a zoologist's delight of animal superlatives. Whales include the largest species ever (blue whale), the largest-toothed predator ever (sperm whale) with the longest intestine (three hundred meters [a thousand feet]) and largest brain ever—some five times larger than our brain. Bowhead whales have the largest mouths ever, and it goes without saying, a whale, the right whale, has the largest testes ever (five hundred kilograms [eleven hundred pounds]).

As whales are by far the largest mammal species, as some are the largest animal species of all time, and as large size and long life are generally related, whales should hold a special place in our consideration of why, where, and how exceptional longevity evolved.

As we have seen, determining the longevity of animals in the wild holds many challenges. Those challenges are multiplied when species live in habitats that are difficult to access, are rare or highly mobile, or live long relative to humans. Whales and dolphins display all of these challenges. The most straightforward and reliable way to establish species longevity in the wild is to identify individuals early in life when the date of their birth is clear and to monitor them until they die. There are a just a handful of cetacean species in just a handful of places worldwide where researchers have been able to do just that.

BOTTLENOSE DOLPHINS

One of those species is the common bottlenose dolphin (*Tursiops truncatus*). One of those places is Sarasota Bay on the central western coast of Florida.

Dolphins have been embedded in human consciousness for more than three millennia. They were often depicted in Greek art as rescuers of humans, possibly because as dolphin trainers discovered two thousand years later, they like to push things (including humans) up to and through the water's surface with their snouts. This is likely due to the instinctive habit that female dolphins have of pushing their newborn calves to the surface in a similar way for their first breath of air. The skeptic in me has always considered these classic tales to be what scientists call *ascertainment bias*. Only drowning sailors who happened to be pushed upward toward safety survived to tell their stories. Those pushed in another direction left no tales.

Dolphins are highly intelligent, with brains bigger than our own. They pass the mirror test of self-awareness. They communicate with one another in what may be a crude language. Because they are highly trainable, dolphins are one of the few animals to star as heroes of their own television series. I'm thinking of *Flipper*, which ran from 1964 to 1967 and which can no doubt still be seen on some cable station at some time of day or night. There were also two *Flipper* movies (1963 and 1996). Maybe the most memorable animal experience of my childhood was my shock and awe at seeing a trained dolphin in a show at Marineland of Florida explode out of the water to grab a fish dangling several *stories* above its pool.

We know far more about dolphins that any other cetacean. This species is relatively small, averaging about only three hundred kilograms (660 pounds)—about the same size as a zebra. They live in warm and temperate seas and along coasts worldwide. Along the coasts of North America, they can be found in water as cold as 10°C (50°F) and as warm as a bathlike 32°C (90°F). Although dolphins do fine in shallow water along coasts, where they can be most easily studied, they can also be found in the open ocean, where they can dive as much as a thousand meters (3,300 feet) deep if necessary, using echolocation to find fish, squid, or shrimp from the pitch-black sea bottom. To do this, they rely on bursts of swimming speed up to thirty-two kilometers (twenty miles) per hour. As big as they are, they are still subject to attack from other predators, and about half of all adults bear scars from shark attacks. The fact that so many have survived shark attacks suggests that they are reasonably successful at fighting them off.

Randy Wells, now director of the Sarasota Dolphin Research Program, has been studying the Sarasota Bay dolphins since 1970, when he was a high school student. That thirty-kilometer (eighteen-mile) long bay, eight kilometers (five miles) at its widest, is bordered by the towns of Sarasota and Bradenton, which combined have more than a hundred thousand people. The bay is only three to four meters (ten to fourteen feet) deep in most places, pleasantly swimmable (by humans), as the water temperature ranges seasonally from 20°C to 30°C (69°F to 86°F). It and the surrounding sea are permanent home to more than 150 dolphins. One of the things Wells discovered by identifying individual dolphins in his early years of study was that more than 90 percent of the ones he saw were permanent residents of the area and so could be monitored year after year. Having by now monitored six generations and thousands of dolphins in that bay, Wells has learned many of the intricacies of the complex social groups in which they live. Those social intricacies may be what led to the development of their relatively large brains. Dolphin encephalization quotient is about 3.3, a bigger brain for their body size than most monkeys. Wells discovered that females in that population give birth for the first time at around eight to ten years of age and that calves suckle for up to two years and are tended by their mothers for as long as six years. He and his many colleagues also found that males can sire calves when they are as young as ten years old. He also discovered that even a fifty-year study is not long enough to follow the longest-living dolphins from birth to old age and death.[1]

It is possible to determine the minimum longevity of an animal that lives longer than it has been studied, as we have seen in elephants and chimpanzees. In the case of dolphins, for instance, if a female were spotted swimming with her calf at the beginning of a fifty-year study, then we could estimate that she is at least calf-bearing age (the youngest calf-bearing age is about eight years), and if we then see her for another fifty years, then she would be at least fifty-eight years old, possibly considerably older as female dolphins reproduce into their forties. That is not, however, how the longevity of Wells's dolphin FB15, aka Nicklo, the longest-lived wild dolphin known to date, was established. It was established with her teeth.

The ages of terrestrial mammals are sometimes estimated by tooth wear. Over the years, teeth are worn down by grinding abrasive food and grinding against one another when chewing. As we've seen, elephant longevity may be limited by tooth wear. However, tooth wear is highly variable, yielding only the vaguest of estimates, little more than young versus middle-aged versus old. Dolphin age is not estimated by tooth wear but by growth lines in their teeth. Such lines were first noticed in the middle of the nineteenth century, but their use in precisely aging dolphins was validated only in the 1980s using known-age dolphins from Wells's Sarasota Bay study. One of the unique features of that study is that some of the animals are routinely captured in shallow water using nets. They are brought aboard a specially designed veterinary examination boat, where they are weighed, measured, given a thorough veterinary work-up, and released back into the bay. Sometimes, early on, a tooth was removed during that examination to determine the dolphin's age. The process is not as drastic as you might imagine. Dolphins have lots of (up to a hundred) small teeth, so sacrificing one—number 15 in the lower left jaw, to be exact—in the service of science seems worth it, especially when that science is focused on conservation of the species in question. There has never been any evidence that doing so has any lasting effect on the animals.

As with determining the ages of clams with shell growth lines or sharks with vertebral growth lines, the older an animal is, the slower the growth and closer together are the lines, making them more difficult to count and more subject to error, which is why I am happy to report that Nicklo's birth date was established in 1984 when she was not particularly old. She was a middle-aged thirty-four-year-old mother. During the next thirty-three years, she was spotted again and again, more than eight hundred times in all, recognized photographically like all the Sarasota Bay dolphins by nicks, scratches, and other marks on her dorsal fin. She successfully bore at least four calves, including Eve, born when Nicko was forty-eight years old. Nicklo was last seen in 2017, when she presumably died or was killed at age sixty-seven years.[2]

I should note that Nicklo was exceptional for her species—exceptional like the people who live a hundred years are exceptional. Female dolphins in this population rarely live beyond age fifty, and males seldom beyond age forty. The oldest male recorded in Sarasota Bay to date, by the way, lived fifty-two years. Using Nicklo's age, we can calculate a longevity quotient for this species and this population at about 2.2, which is a greater LQ than any terrestrial mammal its size or larger and about the same as most monkeys. Note I say "this species and this population" because these dolphins living in and around a shallow, warm-water bay bordered by dense human populations with potentially disturbing human water activities—such as fishing, water skiing, jet skiing, and the pollution from boat engine exhaust, spilled gasoline, and city runoff—may not be typical of the species as a whole. They could be shorter-lived or even longer-lived than dolphins living elsewhere or in the open ocean. Remember that *Arctica* clams' longevity varied dramatically between different populations.

On the other hand, they may be more representative than we think. There is, in fact, another long-term, ongoing dolphin study in a much more pristine area. Shark Bay in Australia is a World Heritage site named for its enormous, shallow, warm water bay, with its large Peron peninsula looking somewhat like a finger making an obscene gesture thrusting straight into it. Shark Bay is about midway up the largely deserted western Australian coast. It is most notable for its unique fauna, particularly on its islands, which have been turned into refuges for endangered species, such as the Shark Bay mouse, the Shark Bay bandicoot, the banded hare-wallaby, the rufous hare-wallaby, and, of course, the boody (*Bettongia lesueur*), with its other, possibly more descriptive name, Lesueur's rat-kangaroo. It does look somewhat like an overgrown rat, although it is a marsupial. We are in Australia, after all.

The bay includes a rich resident population of dolphins, gets seasonal visits by humpback whales, and is home to an eighth of the world's dugongs. Dugongs, by the way, are easily confused with manatees, their closest evolutionary relatives. Both are seagoing, seagrass-eating mammals descended from a common ancestor with elephants. Manatees, to complete the parallel, are common in Sarasota Bay.

At 160 kilometers (a hundred miles) long and eighty kilometers (fifty miles) wide, Shark Bay is considerably larger than Sarasota Bay. The average depth of Shark Bay—nine meters (thirty feet)—is twice that of Sarasota Bay but with roughly similar tepid water temperature. One enormous difference from Sarasota Bay is that Shark Bay has virtually no people living around it. What it does have is plenty of dolphins, three hundred to four hundred at any one time. In fact, it has been known from the 1960s for its tame dolphins that splash in shallow water along the shore, eagerly accepting handouts from the few human beachgoers. The ability to observe these tame, beach-going dolphins close up gained the attention of American behavioral biologists Richard Connor and Rachel Smolker, who camping out on this deserted beach, began studying their behavior in 1982. Two years later, they began the long-term study that continues today.[3]

The Shark Bay dolphins are a different species, the Indo-Pacific bottlenose dolphin (*Tursiops aduncus*), than the common bottlenose dolphins, but they are similar enough that they were considered a single species until several decades ago. Both species reach reproductive maturity at about the same age, with about the same length of calf dependency. The Shark Bay dolphins seem to live about as long as their Florida brethren, at least into their late forties or early fifties, although because the study began more recently and because they do not remove teeth of study animals, we don't know if any have survived as exceptionally long as Nicklo. We do know that male dolphins in Shark Bay form small coalitions to help them separate fertile females from the rest of their social group for mating. We also know that a small group of females has invented a unique foraging tool. They rip loose giant sea sponges from the sea bottom and place them over their noses. Admittedly, this looks a bit ridiculous, probably even to other dolphins, but they use the sponge to protect their snouts when they probe the sea bottom to flush partially buried prey. The bottom here is covered with sharp rocks and broken shells. A good sponge is hard to find, apparently, because having to drop it to eat, they will often retrieve the same sponge and reuse it or even take it with them when they move to another foraging area. This behavior is cultural—passed down from mother to daughter—and it takes some years for young dolphins to perfect the technique. Finally, it has been firmly established here that females age reproductively, beginning fairly early in life, something reminiscent of human females, whose physiological reproductive peak is typically reached in their mid to late-twenties and declines thereafter. Reproductive aging in the Shark Bay dolphins can be seen in that calves of older dolphin mothers are less likely to survive than those of younger mothers and the time between births increases as females grow older.[4] Giving birth and rearing a calf—these are mammals, after all, so don't expect the father to provide any help—apparently takes more out of older females.

So dolphins in the wild do show signs of aging, but they still live about twice as long as a terrestrial mammal of the same size that lives a cushy well-protected, well-fed zoo life.

This is more impressive to me than it may seem at first glance to you. Remember dolphins are mammals and like other mammals maintain a body temperature of about 37°C (98.6°F). They do this while living in water that is anywhere from a little—to a lot—colder than their body temperature. As water whisks heat out of the body twenty-seven times faster than air, dolphins lose heat into the ocean much faster than, say, a zebra would to a similar air temperature. This points out an important and stark difference between ectothermic sharks and endothermic dolphins and whales. At cooler water temperatures, shark metabolism slows, but dolphin metabolism increases. Dolphins need to generate enough heat to balance that lost to the water. This is the fundamental physiological problem faced by aquatic mammals, which requires them to be either very large (which limits heat loss due to a favorable surface area to heat-producing weight ratio), be very well-insulated (by, say, dense fur or thick blubber), or have an exceptionally high metabolic rate—or be all of these. Blubber, by the way, is wonderfully useful to aquatic mammals. It provides insulation and buoyancy plus is an efficient way to store energy, as our human bodies know only too well.

Consequently, adult dolphins have a cocoon of about 2.5 centimeters (one inch) of blubber under their skin, enough blubber to comprise about a fifth of their body weight. They also have two to three times the metabolic rate of a terrestrial mammal of the same size, which means they have to eat two to three times as many calories. Also, remember that metabolism generates oxygen radicals and other damaging by-products. So high metabolic rate is generally, if loosely, correlated with shorter life. This is why I find their longevity quotient of greater than 2 particularly impressive.

KILLER WHALES

Another charismatic cetacean species is the aptly named killer whale (*Orcinus orca*). The aptness comes from *killer*, not from *whale*. If bottlenose dolphins are zebra size, killer whales are elephant size. Despite their name, killer whales are dolphins, the biggest species in the dolphin family, in fact. In nature, they eat virtually any animal that is big enough to bother with, and with their exceptional intelligence and ability to hunt cooperatively, they are capable of killing and eating almost anything, including whales much bigger than themselves. Even massive sperm whales, the largest carnivores on earth, are not immune from attack. They have been recorded eating large baleen whales, and sharks—even great whites—avoid them, as do seals, sea lions, walruses, narwhals, on down to penguins and many kinds of fish. Let me be quick to note that killer whales are simply doing what nature has designed them to do.

One species they have not been recorded to eat, I should point out, is humans. There are zero records of fatal attacks on humans by *wild* killer whales. You can't say the same for captive trained whales, which have killed several people—their trainers—sometimes in a very public fashion. These were not predatory attacks, though. They were more like I'm-having-a-really-bad-day-don't-mess-with-me attacks. But when you're as big as a killer whale, your bad day can be fatal to smaller, terrestrial mammals who happen to be close by.

As I say, killer whales are exceptionally charismatic. They play a role in the mythologies of indigenous coastal cultures worldwide although originally held in low esteem in the industrialized world. For instance, they were used as targets for practice bombings by the Royal Canadian Air Force during World War II. They were also accused of depleting commercially valuable salmon stocks. More recently, their public image has dramatically reversed, possibly due to calling them *orcas* (a name that biologists tend to avoid) instead of *killer whales*. The 1993 *Free Willy* movie, about the friendship between a troubled twelve-year-old boy and a troubled we-don't-know-how-old killer whale, made more than $150 million worldwide, leading to three sequels and an ill-advised attempt to release the eponymous movie star back into the wild. Decades of Shamu-type exhibitions at marine parks no doubt contributed, too. Despite some public concern about keeping such large, intelligent species in captivity, their shows at marine parks are highly popular. Killer whale merchandise is a lucrative offshoot.

Regardless of their public image, formerly bad but now good, these are fascinating animals, and several wild populations have been studied in the coastal waters off of Washington state and British Columbia since the early 1970s.[5]

Killer whales can be found from Arctic to Antarctic seas and virtually everywhere in between. They have even been occasionally spotted as far as 160 kilometers (a hundred miles) up large rivers such as the Pacific Northwest's Columbia River. Virtually any water temperature seems to suit them. They have the second-largest brain on the planet after sperm whales and with that large brain live in multigenerational groups called *pods* with complex social interactions. Pods typically consist of an older female and several generations of her descendants of both sexes. They are one of the only mammals in which both sexes remain with the group into which they were born. In most mammals, one sex or the other emigrates by the time they reach reproductive maturity. Like other dolphins, they communicate with a complex series of clicks and whistles. They are capable of extensive learning, including mothers apparently teaching their calves various hunting strategies. Although considered a single species for decades, recent genetic studies suggest they may be in the process of splitting into several distinct species.

As with other dolphins, though, almost all information we have about these wild ones comes from relatively easily studied nonmigratory populations living near the coast. Also, like other dolphins we know, there are other, ocean-going, more mobile populations about which we know very little. The coastal resident killer whales prefer to dine on fish, particularly salmon. Their reputation as salmon hunters was well-earned. Those migrating through specialize on other mammals—seals, sea lions, and the occasional whale. Finally, yes, like other dolphins, a nearly fifty-year study is not long enough to monitor the longest-lived individuals from birth to death.

The late, pioneering killer whale researcher Michael Bigg identified these two whale types—residents versus migrants—early in his British Columbia study. In fact, the wide-ranging migratory, seal-eating ones are now called Bigg's killer whales. Like other dolphins, individual killer whales can be identified by characteristic markings that can be photographed when they surface. In this population, after just the first fifteen years of study with hundreds of observations on specific individuals, researchers had already concluded that females grew until about ten years of age, had their first calf when about fifteen years old, and had their last calf at roughly age forty. Males approached the lower limit of adult size in about eight years but continued to grow until their late teens. They also concluded that males lived around thirty years on average, with the oldest ones reaching fifty to sixty years, whereas females lived around thirty-five years on average but up to eighty to ninety years at the extreme. It was the extreme values, combined with the fact that some sixty-eight killer whales had been live-captured from these populations for marine park exhibitions, that started the disagreement.

Extreme longevity stories about killer whales had been around for quite some time. One excellent yarn was that of "Old Tom" of Twofold Bay, Australia. Twofold Bay, on the southeast corner of Australia, is on a migration route of humpback and right whales moving between their tropical ocean breeding grounds and their Antarctic feeding ground. During the heyday of whaling in the nineteenth century, both of these species were prized for their valuable blubber, which could be turned into oil used as lamp fuel and lubricants—increasingly needed during the Industrial Revolution. Their hard but flexible baleen was also valuable for use as corset hoops, umbrella ribs, buggy whips—things we use plastic or fiberglass for today. There never was a commercial killer whale industry. They were too small, too fast, too intelligent, and too dangerous, plus they had too little blubber and no baleen.

A small shore whaling industry grew up in Twofold Bay, meaning that rather than ply the seas looking for whales, whalers could row out from shore, spot and harpoon a whale, and tow it back to shore for processing. Local killer whales learned to follow these boats, dining on the good parts— the tongue and lips—of the dead or dying baleen whales as they were being towed. Eventually some of the cleverer killer whales learned that if they herded baleen whales into Twofold Bay and alerted whalers to their presence and location, they would soon be rewarded with a fine tongue-and-lip meal. And so there developed a mutually rewarding symbiosis between whaler and killer whale. The whales were even given pet names—Cooper, Humpy, Hooky, Jimmy, Kinchie, Skinner, Stranger, Typee, Walker, Big Ben, Big Jack, and, of course, Little Jack.[6]

Tom, who evidence suggests may have been a female (figure 13.1), helped the whalers especially well. She would reputedly position herself in front of the whaling station, thrashing her tail fluke to alert the whalers, who then followed her to where the rest of the pod had a baleen whale surrounded. Identifiable by some unique marks on her dorsal fin, Tom—or a whale that they consistently thought was Tom—was known by three generations of the whaling Davidson family, who had been active in the bay since 1846. Eventually, Tom became Old Tom. When her body washed ashore on September 17, 1930, it was considered a local tragedy. Calculating from Davidson family lore that Tom must have been at least ninety years old, her skeleton was reverently preserved, and a museum was built to display her bones and to tell the history of the Twofold Bay whaling industry. These bones can still be seen today at the Eden Killer Whale Museum at Twofold Bay. And so the story of the ninety-year-old killer whale passed into history.

As grand as is the story of Old Tom, two whaling experts, Edward Mitchell and Alan Baker, made it their mission to validate—or not—Old Tom's actual age. They scoured the historical records but could find no tangible evidence of Old Tom other than stories handed down through the Davidson family. There was, however, a 1910 home movie in which Tom appeared as a full-grown adult, recognizable by that distinctive dorsal fin. Assuming that

Figure 13.1

Old Tom helping whalers of Twofold Bay. The photo is taken from a 1910 documentary film about Old Tom. The towed baleen whale that has just been harpooned is out of frame. Notice that there is a whale calf between Tom and the whaling boat, suggesting Tom may have been a female. He or she died at the reputed age of more than ninety years old, although her teeth suggest that either the corpse that washed ashore in 1930 was not Old Tom or that it was a succession of different whales that helped the whalers all those years.

it takes fifteen years to reach adult size and knowing that Tom died in 1930, that would make her at least thirty-five years old. How much older than that was she really? Mitchell and Baker visited the Museum in 1977, examining Tom's 6.7-meter (twenty-foot) long skeleton in great detail, including noticing that three of Tom's teeth were missing and two were severely abscessed. They were given permission to take a tooth from Tom's lower jaw and examine its growth lines. From that evidence and the film evidence, they concluded that Tom, against all folklore, was indeed about thirty-five years old.

Whether the story of Old Tom had anything to do with it isn't clear from the early estimates, after only fifteen years of study, that British Columbia killer whales could live up to ninety years. What is clear is that there is considerable disagreement about this estimate. It started when veterinarian Todd Robeck and colleagues from SeaWorld published a life history analysis of their captive killer whales, nineteen that were captured from the wild between 1965 and 1978 and another sixty-five animals that were born in captivity.[7] Their paper compared the growth, reproduction, and survival of captive killer whales with the wild British Columbia whales. They found, not surprisingly, that the well-fed, sedentary animals at SeaWorld grew a bit faster and matured a bit earlier than the wild whales. They also concluded that although their killer whales had survived less well than wild whales prior to the year 2000, as captive husbandry improved after the year 2000, captive survival became at least as good as that of wild whales. Finally, they noted that the oldest known-age animals in the SeaWorld population and the wild—those animals with unquestionable birth records—were less than fifty years. In fact, they said further that only about 3 percent of killer whales were known to survive even that long. If that was the case, then estimates of the ages of very oldest wild ones, which by now had been estimated to over a hundred years, were likely serious overestimates. They concluded that female killer whales probably lived no longer than about seventy years, with males living maybe a decade less.

That conclusion did not sit well with some of the wild killer whale researchers or animal rights activists, who were sure that captive whales must survive less well than wild and free ones.[8] To help understand what followed, we need to make some distinctions about assessing age in wild animals that I've avoided previously but that now become important. Let's call animals with birth years known from direct observation *known-age* animals. *Estimated-age* animals are those with ages that can be estimated to within a few years or a few percentage points of their real age because they were first seen when partially grown. Their age can be back-calculated from the sizes of young known-age animals. We have already done this for the oldest chimpanzees and elephants. Then there are *guesstimated-age* animals. Their age is based on a variety of assumptions that, if violated, could make that age dramatically wrong—off by maybe 100 percent or even more. Greenland shark ages are guesstimated. This is not to criticize such guesstimates but only to acknowledge their uncertainty. Researchers do the best they can with the information at hand.

Let me illustrate the magnitude of uncertainty with an imaginary example. Say that a group of three female killer whales is spotted for the first time. One is a nursing calf, and another is the nursing mother (let's call her NM for "nursing mother") that she suckles from. The third is still not fully grown but close. Call her Juvie. Because we know that killer whales live in multigenerational family groups, we can assume they are all related, likely a mother and her two daughters. The age of the calf is unequivocal because of its size and the fact that it is still nursing. Call it one-year-old. Juvie, we might reasonably assume, is the older sister of the nursing calf. Because we know that females in this population can have calves as often as every five years and because of her size, we can estimate Juvie is about six years old, give or take at most a year or two.

Now we start guesstimating. NM has had at least two calves. If these were her first two calves, she could be as young as twenty-one years old, back-calculating that first births occur at the average age of fifteen plus five years between births plus one year for the age of the nursing calf. Of course, she could have lost her first calf or two. So these could be later calves, and she could be considerably older. As we know that killer whale females seldom reproduce after age forty, NM could be as old as forty-one or even older. One forty-six-year-old has been documented to give birth, so she could be forty-seven at the extreme. So depending on our assumptions, all of which are reasonable, NM is somewhere between twenty-one and forty-seven years old—not exactly precise, which is why I call these guesstimates.

I give this long example to point out how guesstimates based on various reasonable assumptions can be wildly different. The longer a study goes on, the fewer guesstimates need to be made, and the more known-age and estimated-age animals there will be.

Which brings us back to the disagreement between the SeaWorld and British Columbia killer whale researchers. SeaWorld whales were mostly known-age because most were born in captivity. They made no claim to have really old ones. The presumed oldest wild British Columbia whales were guesstimated-age animals from early in their study. The longest-lived, according to those early guesstimates, was a whale named J2 in the scientific literature and "Granny" in popular accounts. Granny was photo-identified in 1971 in the company of a fully grown male known as J1 or more fondly "Ruffles." It was assumed from their behavior that Ruffles was Granny's son. As she had no other calves over the next sixteen years, the researchers assumed that she was postreproductive and that Ruffles was her last calf. The guesstimated age then went like this. Males reach full adult size at about twenty years of age, and females seldom reproduce past the age of forty, so if Ruffles was twenty and Granny was postreproductive, then she could have been born at least sixty years earlier—in 1911. If she was sixty in 1971, then she would have been seventy-six in 1987 when they began estimating the ages of their study animals, and when she was last seen in 2016, Granny could have lived 105 years.

The remote probability that an animal lives more than a century when only 3 percent of that species lives to age fifty is only part of the problem with Granny's guesstimated age. That, by the way, would be similar to a person living in preindustrial times living to at least 150 years, which is roughly twice the age to which 3 percent of those populations survived. The other problem is that Ruffles turned out from later genetic analysis not to be Granny's son after all.[9] Without any offspring (and no offspring of Granny's appeared in later genetic analyses), we are left with the fact that Granny did not leave any surviving calves after 1971. So if she were really postreproductive at that time—and was not infertile for some reason or had lost all her calves due to environmental hazards—then it would be reasonable to assume that she was at least forty years old in 1971 and therefore born in 1931 and at least eighty-five years old when last seen. On the other hand, if she were infertile or her calves had all perished before they were noticed (killer whale calves are frequently not noticed until they are about six months of age), because none of the whales in her pod of about twenty-five animals were genetically determined to be her calves, then she could have been as young as twenty years old in 1971 (because she was full adult size at that time), making her around sixty-five years old when she disappeared. I have to say that this last estimate is highly unlikely given that she seemed to have a leadership role in her pod early on and leadership roles are usually filled by the oldest female in a pod.

So the maximum age that killer whales can reach remains something of a mystery. A reasonable guesstimate from my perspective is to conjecture that Granny was around forty years old in 1971 and so around eighty-five years old when she finally disappeared for good. That gives killer whales a longevity quotient of 1.6. For her size, she was longer-lived than an average five-thousand-kilogram (eleven-thousand-pound) mammal but not as long-lived most primates. That is a bit longer than the estimated age of the oldest wild African elephant, a species close to the same size.

Researchers who study them in the wild still assert, despite the questions about Granny's age and the lack of direct evidence, that females can live eighty to ninety years, which seems reasonable. SeaWorld researchers continue to estimate that sixty to seventy years, conceivably up to seventy-five years, is much more likely. They both agree that females reproduce only to about age forty. So whether they can survive as long as twenty to thirty-five additional years or forty to fifty additional years, they should still have substantial postreproductive lives, as do women. Women become physiologically incapable of reproducing by about age fifty, when they had a remaining life expectancy of several decades even in preindustrial times. So could killer whales be an interesting and informative example of something like human menopause? Menopause is an evolutionary puzzle that we will take up in the next chapter.

BALEEN WHALES

Evolution eliminated the teeth of fifteen species of whales. Instead of biting and tearing their prey, these species feed by gulping enormous mouthfuls of water and sieving out small prey like shrimp and krill through what has replaced their teeth—baleen. None of these species has the convenient behavior of remaining year-round in one relatively shallow bay or coastal waters so that they can be closely and continuously monitored decade after decade. As they have no teeth to saw into sections to count growth lines, do we know anything about how long they live and, if so, how?

These are the largest of all whale species, the ones that were hunted most vigorously during the heyday of whaling in the late nineteenth century. The blue whale, largest of them all, weighs more than thirty killer whales. Although baleen whales occur throughout the world's oceans, they spend considerable time in either north and south polar seas, where their size and thick blubber protect them from hypothermia in those frigid waters. Indigenous people of the Arctic hunted whales long before the whalers of the nineteenth century, eating their meat, turning their blubber into oil, and using their baleen to fashion baskets, snares, toboggan runners, and a variety of other useful tools for Arctic living.

Most large baleen whales migrate seasonally between warm and cool waters. They feed during the summer in cold polar oceans where krill and other small prey bloom in profusion. In the fall and winter, they migrate to warmer waters, offshore of Florida or Hawaii or Baja California, where they breed and seldom, if ever, feed. The relatively slow metabolism allowed by their great size and their abundant energy stored as blubber allow them to go many months without eating. It is this separation between feeding and breeding seasons, between periods of energy storage and energy depletion, that has allowed us to learn something about how long they live. Yes, as you may have suspected, those seasonal cycles do affect their earwax production.

They are called *earplugs*. They largely consist of earwax, the technical name of which is cerumen, a name you can now forget because earwax is nicely descriptive. Whale auditory canals have a mere slit of an opening to the external world, so unlike our own earwax, which eventually is pushed out of our ears, whale earwax does not escape. It accumulates throughout life, forming seasonal dark and light growth lines, the color depending on whether they were feeding or not when the wax was produced. Seasonal ear wax production captures many aspects of an individual whale's life. In addition to age, chemical examination of ear plugs has revealed hormonal profiles of when the wax was formed as well as of exposure to environmental pollutants. Even the sea temperature and the food they were eating at different periods of their lives can be reconstructed from earwax layers.

The main human use of earplugs for decades, though, has been for estimating baleen whales' ages.[10] The main drawback, needless to say, is that earplugs are available only from dead whales. As whales are now internationally protected, we are left with only historically collected samples or those from the carcasses of newly stranded animals or from the small number that indigenous people have permission to hunt. We have learned from earplugs that the largest whales live a long time—not clam- or tubeworm-long but at least as long as the longest-living humans. Blue whales, for instance, have been documented from earplug analysis to live at least 110 years, and fin whales at least 114 years. Because of their size, though, more than a hundred times the size of a person, these ages represent less than overwhelming longevity quotients of around 1.0 to 1.5. Of the baleen whales, the one that deserves special mention for its longevity is the bowhead whale, which seems to be the cetacean longevity champion.

BOWHEAD WHALE

Bowhead whales (*Balaena mysticetus*) do not migrate to warm waters to breed. They live in the Arctic and subarctic year-round, although they migrate considerable distances east and west and a bit north and south as the sea ice waxes and wanes. At about sixteen meters (fifty-two feet) and seventy thousand kilograms (seventy-seven American tons), they are the second-heaviest whale. They are the most northerly whale species known, and they spend their lives in close-to-freezing water, reminiscent of the Greenland shark. Another common name for bowhead whales is, in fact, the Greenland whale. The "bowhead" name comes from a massive head, which they use to batter through as much as sixty-centimeter (two-foot) thick Arctic sea ice. They take advantage of this ability to escape from their only predator, killer whales, by swimming under thick ice that only they can break through to breathe. As might be expected from living in such cold water, they have an exceptionally thick coat of insulating blubber (fifty centimeters or about twenty inches), so thick, in fact, that like the right whale, they float after death. Bowheads were severely depleted by commercial whaling that lasted into the early twentieth century, although they have rebounded nicely in recent times. They have been protected by international law since the 1970s. These days, they are hunted only by special permission given to indigenous peoples of the Arctic.

Unlike species that migrate between warm and cold waters, bowheads, alas, do not form growth lines in their earplugs. Repeated photographic "captures" of the same few bowheads year after year and examination of the carcasses of hunted whales indicate that they reach puberty when about thirteen meters (forty-three feet) in length, which puts them somewhere between their late teens to mid-twenties. Such late puberty suggests that they may be exceptionally long-lived. That suggestion was strengthened in the 1980s and 1990s, when a few killed bowheads were found to have old harpoon points made of ivory, slate, and stone buried in their blubber. By matching these points with those from anthropological artifacts in the Smithsonian Museum, researchers determined that such tips had not been used in at least a century. So the whales have been alive for at least that length of time.[11]

Whale researcher John "Craig" George from the Alaska Department of Wildlife Management wanted to make better guesstimates for bowheads than "at least a century" but had no earplug or tooth growth lines to estimate a more precise age, what was he to do?

Very sensibly, he contacted chemist Jeffrey Bada of Scripps Institution of Oceanography in San Diego, California. Bada's main research interest was in the chemistry of life's origin on earth and possibly elsewhere. His specialty is amino acids, the building blocks of proteins. Proteins are hallmarks of living organisms. He had examined amino acids found in the deep ocean around hydrothermal vents and those that arrived on earth on meteorites. He had helped design an instrument package to search for life on Mars. An offshoot of that work had been Bada's pioneering research on the use of something called *amino acid racemization* to estimate the age of proteins.

Despite the technical sounding name, in principle, the technique is straightforward. Like baseball pitchers, amino acids can exist in either left- or right-handed forms. Nearly all living organisms produce only left-handed forms, a quirk of nature discovered by Louis Pasteur in the nineteenth century when he wasn't busy discovering the rabies vaccine or the fact that germs caused disease. Over time, the left-handed forms spontaneously convert to the right-handed forms. That is the racemization process. Racemization occurs at a known rate that differs for each of our twenty different amino acids, so by measuring the ratio of left- to right-handed forms, it is possible to estimate how much time has passed since any protein of interest was first formed.

Bada had used amino acid racemization to estimate the age of everything from ancient human bones to ocean sediments. It could even be used to assess the age of proteins on a scale relevant to, say, a human or whale lifetime. This was established by comparing racemization age to calendar age in human teeth, which, thankfully, were not extracted for the purpose but collected from dentists and patient records. For the same reason that the cores of eye lenses were used to guesstimate the ages of Greenland sharks—lens cores are formed *in utero* and persist as long as an animal lives—they might be used to estimate the age of bowhead whales. So George provided Bada with the eye lenses of forty-two bowhead whales that had been legally killed by indigenous hunters to see what they revealed about whale age. Comfortingly, the results were consistent with other estimates that bowheads reached puberty in their mid-twenties. The lenses also revealed that bowheads appeared to continue growing until they approached age fifty. Then came the huge surprise. Although most of the adults in this sample were estimated to be between twenty and seventy years old when killed, three, all males, were estimated from racemization to be more than 150 years old. The oldest was judged to be 211 years old, nearly twice as long a life as any other whale species.[12]

Let me note right away that whale researchers themselves interpreted these guesstimates very conservatively, concluding that "longevity in excess of 100 years is not improbable." Aging researchers like me, however, always on the lookout for examples of exceptional longevity, seized on this 211-year estimate as gospel truth, and it has been commonplace in the field ever since the publication of that paper in 1999. I have stated unequivocally in aging conferences ever since that the longest-lived mammal is the bowhead whale, which can live more than two centuries. That is a reasonable longevity quotient for an endothermic, Arctic-living whale of 2.6. But can it really live this long?

Here is the complicating issue, though. The rate of amino acid racemization depends on temperature. Like many chemical and biological processes, it proceeds slower when colder and faster when warmer. So any use of racemization to estimate age has to make assumptions about the temperature that the proteins experienced since their original synthesis. If eye lens temperature during a whale's life was lower than assumed, the age of the whales would be *under*estimated. As all mammals have roughly the same internal body temperature, 37°C (98.6°F), the temperature complication wasn't much of an issue for teeth. The inside of the mouth, after all, is where we often measure body temperature when other orifices are off limits. But what about the eye lens?

Bowheads, spending their lives in frigid Arctic waters, have a somewhat lower than normal mammalian body temperature. For the original study, published before anyone had measured internal bowhead temperature, the age calculations conservatively assumed it was a fraction of a degree lower than most mammals or halfway between what was known about human and fin whale internal temperatures. By the time a further forty-one bowhead eye lenses were analyzed fourteen years later, bowhead core body temperature had been found to be more than 3°C (5.4°F) lower than most mammals, and the estimates were changed accordingly. Of the thirteen adult whales in this sample, once again, most were relatively young, between twenty and ninety years, but one male was judged to be 146 years old. Incidentally, using actual measured bowhead body temperature to reevaluate the original age estimates, that 211-year-old male would be estimated at more than 250 years old—something, I hasten to add, that I calculated, not something that the bowhead researchers claimed.

One other thing makes bowhead longevity more astonishing than ever. A recent study has discovered that baleen whales in general have much higher metabolic rates than previously calculated from their size alone. The researchers didn't measure metabolic rate itself, which would be virtually impossible in a free-living creature so large; instead, they used sophisticated new methods to calculate how much giant whales eat. It turns out that baleen whales, including the bowhead, eat up to three times as much as previously suspected.[13] This means that bowhead metabolic rate has to be three times higher, and damaging free radicals three times higher, than expected too.

Here are two final thoughts about bowhead whale longevity. First, I suspect that 211 years or 250 years, whichever of these estimates you prefer, may be serious underestimates of the oldest bowhead's real age. I say this because the conservative assumption researchers have made is that eye lens temperature is approximately equal to the bowhead's internal temperature. It is difficult for me to accept that a bowhead's eye lens, which lives just centimeters from nearly freezing water, maintains a temperature anywhere close to the whale's internal temperature. For once, researchers of animals in Methuselah's Zoo are being much too conservative in their guesstimates. Second, I previously expressed skepticism of animal ages that were extreme outliers within their species. For bowheads, I am much less skeptical because of the extensive exploitation of that species by commercial whalers into the early decades of the twentieth century. Only a few survivors from before commercial whaling ceased is exactly what you might expect. Also, for most species, we have documented ages or age estimates or guesstimates for hundreds to thousands of animals. For the bowheads, we have age estimates for fewer than one hundred adults, so age gaps would be expected. This is one species that may be considerably more exceptional in its longevity than we currently suspect. Having long life plus enormous body size, of course, suggests exceptional cancer resistance in bowheads and other large whales. Interestingly, some recent investigations of whales' genomes have suggested that they have some special tumor-suppressor genes that would be worth a closer look by cancer biologists.[14]

IV HUMAN LONGEVITY

There must have been a final day that an adventurous troop of chimpanzees peered back into the safety of the forest they knew so well and made up their minds to remain permanently on the open African savanna. There was abundant food on the savanna. There was also abundant danger and few trees in which to escape ground-bound predators. A sign of their irreversible commitment to savanna life may have been when they finally decided to stop sleeping in the safety of trees.

These were not the chimpanzees of today but an ancestor of today's three chimpanzee species—common chimpanzees, bonobos, and humans. We are not exactly sure what they looked like. They may have been unusual chimpanzees even for those times. Perhaps they had undergone some unique genetic changes to better suit them for savanna life. Maybe they were forced onto the savanna by a shrinking forest as the climate dried. All we know is that the descendants of those chimpanzees were wildly successful over the next 6 million years. They diversified into many species and spread throughout Africa. By 4 to 5 million years ago, they had abandoned their ancestral knuckle-walking and evolved an upright posture. They now walked, trotted, and ran on two legs, possibly because that locomotion mode was more energetically efficient on the open savanna, possibly because it freed up their arms and hands for other activities while they walked or ran. Perhaps both.

By 2 million years ago, their brains had begun the enlargement that would ultimately lead to our brain, three times the size of the other chimpanzees. Some early humans migrated out of Africa, perhaps like the earlier move onto the savanna, forced by climate change and perhaps out of a sense of adventure or a search for new challenges. In fact, several human species would march out of Africa over the next 2 million years, including our own, which, so far as we know today, left long after the others. Today, we think that *Homo sapiens*, our species, left for the first time around a hundred thousand years ago.

I should preface any statement about these early *hominins*, a term coined to distinguish our ancestors taxonomically from the other chimpanzees, by the statement "so far as we know today" because seemingly every month brings new discoveries about early human evolution. Until fairly recently, we thought that anatomically modern humans arose only about a hundred thousand years ago, then new fossil discoveries pushed that estimate back to 200,000, and now 300,000 years.[1] At that time, there may have been as many as nine species of hominins living somewhere on earth.

Along the way, these various species of hominins developed an array of stone tools and mastered the use of fire for cooking and also perhaps for opening up the landscape for easier hunting. The one species besides ourselves that we know a considerable amount about—the Neanderthals—learned how to make rope, weave clothes, fashion jewelry, store food, play music, and paint figures of the animals they hunted on cave walls. We don't know whether they developed language of sufficient sophistication to tell one another stories around the cooking fire at night, but it is easily conceivable.

Leaving Africa, modern humans encountered earlier hominin emigrants such as the Neanderthals. By now these non-African hominins had diversified into at least four species and had colonized much of the globe. We don't know exactly what happened during those encounters, except that we occasionally mated with at least two of these species, the Neanderthals and the Denisovans. We know this because remnants of those matings survive in our genes today. Thanks to our almost miraculous ability to recover ancient DNA from the bones and teeth of these other human species and our development of inexpensive DNA sequencing technology, I have learned, for instance, that about 2 percent of my own genes originated in the Neanderthals. Another thing we know is that by thirty thousand years ago, we were the only hominin species left. Did we exterminate our competitors, or did nature do the job for us?

Until about fifty thousand years ago, there is little reason to think that the length of our lives was much different than that of the Neanderthals, which we know (because we have discovered hundreds of Neanderthal skeletal remains) was nasty, brutish, and short. Estimates for our doomed Neanderthal cousins are that about 80 percent of those who managed to reach age twenty were dead by age forty and the odds of reaching age twenty appear to be no better than fifty-fifty.[2] When I say nasty and brutish, I mean it literally. Nine of ten early human skeletons show signs of major trauma, although whether it was caused by ourselves or by large animal attacks is disputed. One analysis of over a hundred remains concluded that about three-quarters of our ancient relatives showed signs of attack by large carnivores such as bears, wolves, and large cats. Apparently our killing of ancient carnivores was not a one-way street. Competition for prey and shelter must have been intense.

Around forty thousand years ago, when we had spread throughout most of the globe except for the Americas and when Neanderthals and other hominin species were on the decline or gone, suddenly the age mixture of skeletal remains changes.[3] "Older adults" became more common relative to "younger adults." This may have been due to a change in human biology or some cultural developments. We are unlikely to ever know. It did coincide with a seeming increase in cultural complexity, such as the development of things like elaborate body adornments. For whatever reason, about forty thousand years ago, when we see no other obvious change in our culture or biology, it seems like we had begun to outlive our Neanderthal cousins.

There are two stories to tell about anatomically modern human longevity. The first is about how long we lived in what might be called a state of nature before we clustered into cities and towns, before we paid other people to grow our food, make our clothes, fashion our art, and before we were subject to global pandemics. This story is about how long we lived for most of our species' existence. It is a story about our biological design, shaped by hundreds of thousands of years' of interactions with the changing environments through which we lived. This story allows us to compare our biology of aging with other species still living in the wild. The other story is about how long we live now after we have domesticated ourselves—molded our environment to suit our wants if not our needs, developed science, and discovered hygiene and ways to manage our health. That change has been dramatic, particularly in the last century or two. For the last 150 years, life expectancy in technologically advanced countries has increased by a jaw-dropping two and a half years per decade—that is, six hours per day if you're keeping count.[4]

OUR "NATURAL" LONGEVITY

For the great majority of our three-hundred-millennium history, the only hard evidence of how long we lived is what can be estimated from skeletal remains. When police officers find the long-buried corpse of a murder victim with little remaining flesh on its bones, their report usually specifies the sex and age range for the victim. Sex is determined mostly from pelvic shape, although these days it is more often from DNA analysis. Age is determined (or I should say, *estimated*) from features of the victim's size, teeth, skull, pelvis, ribs, joints, and perhaps spinal column compared with skeletons of known-age individuals. Like most techniques of age estimation, this one works well for children, adolescents, and young adults but becomes increasingly less accurate with age. For estimating anthropological remains, which may be thousands or tens of thousands of years old, there is the additional problem of how time, postmortem damage, and deterioration have affected the bones, which is why in the Neanderthal studies, estimates were simple categories—newborn, child, juvenile, adolescent, young adult (those at least twenty years old), and older adult (older than forty). One exceptionally large study of more than thirteen hundred remains of indigenous Americans who lived around a thousand years ago suggest that few, if any, people at that time lived beyond age fifty or fifty-five.[5] A few anthropologists believe fifty to fifty-five years represent the oldest humans of the time. This is hard to credit as we now know that chimpanzees can occasionally live longer than that in the wild. More likely, these age limits are imposed by the story ancient bones can tell. The bones of a fifty-five-year-old or an eighty-year-old ancient human may not be distinguishable.

To understand more about the length of ancient human life during those long millennia in which our modern biology evolved, of necessity some anthropologists study modern humans living in something resembling ancient conditions.

For decades, anthropologists have sought out remnant human populations living under conditions not entirely different than what must have been common prior to widespread agriculture. It is tempting when gazing into the ancient past to be typological—to assume that there was a "typical" human existence—forgetting that forty thousand years ago, modern humans were living throughout Africa, Europe, and Asia (everywhere except the Americas). So at that time, we lived in warm and wet as well as cool and dry forests, We also lived in grasslands, on the frigid Siberian tundra and in parched deserts. These were places where food was abundant, the climate was mild, and life was comparatively easy as well as places where nature's larder offered meager supplies and the climate could be unforgiving.

Two things need to be kept in mind as we sort through what we can learn about the biology of longevity from the study of modern humans living in quasi-ancient conditions. First, by the time scientists study them, even the most isolated people have been influenced by modern life in some ways that we can see and other in ways we can only guess. Missionaries or traders generally beat anthropologists to these populations, bringing ideas, goods, implements, and sometimes diseases with them. Infectious diseases can quickly ravage naive populations, massively distorting the predisease age structure, particularly soon after initial anthropological contact, the period of most interest. That was spectacularly evident during the European colonization of the Americas in the sixteenth century, when according to some researchers, as many as 90 percent of indigenous people died from introduced diseases.[6] Even when missionaries don't arrive first, goods and implements often do through brief and occasional contacts with other groups living in adjoining areas. Second, people living in these remote regions today are living in marginal habitats that modern societies have chosen to ignore, probably for a reason. In ancient times, most people would have lived in more hospitable areas.

Let me admit straight off that my personal interpretations of these anthropological studies are strongly colored by the months I spent living with the Miyanmin people of Papua New Guinea. My former graduate student Keyt Fischer spent some years living there. We were interested in the lives of the animals called *cuscuses* and the tree kangaroos that occupied what in other rain forests of the world would be a monkey's ecological niche—the fruit- and leaf-eating, tree-living, mammalian niche. Although covered by rain forest, primates other than humans had never spread to the island of New Guinea. We were curious whether these marsupial monkey equivalents had, like monkeys, evolved slow development, slow reproduction, and relatively long life. Neither of us was particularly focused on the people themselves. Anthropologists had already studied them. We simply needed the Miyanmin people's help for our studies. Actually, we needed their help just to survive.

Although coastal New Guinea had been known by European sailors since the sixteenth century, most of its mountainous interior remained a mystery to the outside world until well into the twentieth century. Even today, New Guinea's interior has few roads and is mostly accessible only by bush plane.

The Miyanmin live in isolated valleys of New Guinea's central cordillera near the geographic center of the island. They cleared land for a small airstrip in the 1960s, hoping that it might magically bring wealth from outside. When it was finished, people from much of the surrounding area relocated to be near it. The wealth didn't materialize, but it did bring anthropologist George Morren, who first told us about the Miyanmin and the possibility of finding thriving cuscus populations in the nearby mountains, which had been depopulated with the clearing of the airstrip. When Keyt and I visited in the early 1990s, there were roughly a couple of hundred people living near the airstrip, tending small subsistence "gardens" of sweet potato, taro, and bananas. A few semidomesticated and highly valued pigs provided most of the protein. Village dogs lived on scraps and helped with hunting. Men still hunted with bamboo spears and arrows, although women who encountered rats and small marsupials in their gardens efficiently dispatched them with sticks and seemed to provide about as much meat as the men with their hunting forays into the forest. We hired a dozen or so Miyanmin to guide us into the higher mountain forests, where the cuscuses had not been hunted out. It was too far from the airstrip for regular hunting trips. Observing these men—women were mostly forbidden from working with us—during our weeks-long treks through the mountains was a humbling experience. Our high-tech, lightweight tent never kept us as dry as the branch-and-frond shelters our workers fashioned for themselves each night. They climbed trees so much faster using sticks and vines than we could with our fancy ropes and slings that we soon gave up even trying to do it ourselves. Their ability to read the forest was stunning. One night, I became hopelessly disoriented at dusk when radio tracking a cuscus and could not find my way back to camp. After crashing about in every possible direction in a desperate attempt to find a place I recognized, I settled down glumly in the shelter of giant tree roots in the sinking hope that I might find my way in daylight. Keyt became worried when I didn't return. The small search party she sent to look for me found me just before midnight. They had tracked me through the forest *in the dark*. For the next week or so that we explored that area, our guides would sometimes pause as we searched for cuscuses, look around a bit, and then say "you came through here that night you were lost." Clumsy, desperate scientists apparently leave telltale signs in the forest.

As I was only just becoming interested in the biology of aging at the time, I gave little thought to how long the Miyanmin lived, although in retrospect I would have guessed that with malaria, tuberculosis, hunting accidents, personal violence, and a bestiary of intestinal worms and other parasites, it wasn't long. Numbers and counting were nonexistent in their culture, so there was no point in simply asking. My one attempt at deciphering someone's age was when I asked Kwekiap—our closest friend, the best cuscus catcher, and one of the older men in the village—whether he had ever eaten another person. The Miyanmin had a reputation in the past for occasional hunting raids on other villages. Protein had always been in short supply in that part of New Guinea, and according to anthropologist Morren, such raids lasted at least into the 1950s. The question amused

Kwek, who chuckled and, deftly avoiding a definitive answer, said, "Don't worry, I've never eaten a white person."

We were under no illusion that the Miyanmin were untouched by modern life. They may have still hunted with bamboo spears and arrows, but some had highly prized steel machetes. There was a small elementary school along the airstrip, a two-way radio with which they could contact the missionary bush plane that stopped by every few weeks, and even a small store that sold a few items brought in by those bush planes. The one person who spoke a bit of English in the village, Charles, had learned it from listening to the BBC on a short-wave radio. Despite these modern touches, I would have been surprised if anyone in the village was older than about seventy. I would also not be surprised if I were wrong about that. As I say, the people were not my focus.

Like the other species covered in this book, it is worth asking how anthropologists who professionally study aging in these small, isolated, prenumerate populations determine or rather estimate people's age. Generally, they start by working out relative ages—an age chain, so to speak. Questions like "Were you born before or after that person?" or "Who is your oldest (youngest) child (brother, sister)?" or "Which of those people is older (younger)?" can usually be answered by almost everyone in a small village where people have known one another their whole lives. There may be a few inconsistencies, but eventually, you get a pretty good idea of the relative ages of people. Using this method and estimating the actual age of young children or asking whether individual people remember well-documented events (such as the arrival of the first airplane or the year of a widespread natural catastrophe) eventually allows reasonably accurate estimates of everyone's age. "Everyone" in this case means relatively few people, as by definition, these are small, isolated populations. That is not a trivial issue as small samples can contain large biases. The oldest people will be the hardest to assess, but having a reasonable estimate of the age of their oldest child helps, if that oldest child is still alive.

This basic methodology was worked out by Nancy Howell, one of the first anthropologists to focus her attention on how long people lived in these traditional societies. A PhD student at the time, Howell was part of a large Harvard team led by Irven DeVore and Richard Lee that studied a population of !Kung hunter-gatherers living around water holes in southwest Africa's Kalahari Desert in the 1960s. She is quick to point out that the !Kung she studied, like the Miyanmin, should not be seen as examples of pristine human nature. Outsiders had made regular visits to the area since at least 1900. They had neighbors who tended cattle. They were familiar with metal pots, plastic water containers, shoes, and vehicles. Still, they had refused to be "settled" and continued to live almost exclusively by hunting and gathering.

Comparing what we know about the longevity of the !Kung with knowledge of two very different modern hunter-gatherers—the forest-dwelling Ache of eastern Paraguay and the savanna-living Hadza of Tanzania—will allow us to triangulate on what our ancient human life may have looked like.[7] Despite the dramatic differences in the environments they inhabit, there are remarkable similarities among the three groups. In all three, people are relatively small. Men average about 160 centimeters (five feet, three inches), the women about 150 centimeters (four feet, eleven inches) in height. In all three, girls experience their first menstrual period at fifteen to sixteen years old, and all have their first baby at about age nineteen. For comparison, in the United States and Europe today, where physical labor is rare and high-calorie food is not, first menstruation is typically at around twelve to thirteen years. As with other species, people living in the wild, so to speak, where getting one's daily bread requires considerable work, move through life's early signposts a bit more slowly than do their domesticated relatives.

However, there are some clear differences among the groups as well. Life is hardest—meaning survival requires the most work—in the desert-living !Kung. This can be seen from their leanness and low fertility, both reliable signs of what might be called a group's energy stress. A simple measure of leanness—or its opposite—is the body mass index (or BMI), which measures how much someone weighs relative to their height. Current BMI guidelines by the World Health Organization suggest that a BMI between 18.5 and 25 is a normal healthy body weight. That translates to a 175-centimeter (five-foot, nine-inch) man weighing between fifty-seven to seventy-six kilograms (126 to 168 pounds) or a 163 centimeter (five-foot, four-inch) woman weighing between fifty and sixty-six kilograms (110 to 146 pounds). By those standards, !Kung women in the late 1960s had an average BMI of 18.0, making them slightly underweight by modern standards. Their fertility also reflected energy stress. In the 1960s, when Howell studied them, they averaged a very low 4.7 live births over a woman's reproductive lifetime. Putting that number in modern perspective, a contemporary, well-fed, contraceptive-eschewing Mennonite religious group averaged more than twice as many live births per female reproductive lifetime.[8]

Next in terms of energy stress are Tanzania's Hadza, who hunt and gather on the East African savanna not far from the famous, game-rich Ngorongoro crater and luxury tourist camps. Still, some of them have resisted the allure of modern life and maintain a largely traditional lifestyle. Hadza women are less lean (a BMI of 21) and more fertile (6.2 live births per lifetime) than the !Kung. Living in the lap of luxury by hunter-gatherer standards, some Ache still hunted and gathered in Paraguay's forests through the 1970s and 1980s, although other Ache had been enticed on to reservations, where they lived in more modern circumstances. Ache women still living in the forest had an average 24 BMI and eight live births per lifetime.

My assumption is that similarities in longevity found among these hunter-gathering people living in such diverse habitats—with different available food, different dangers from animal and human predators, and different types and degrees of interaction with the outside world—will reflect something important about our longevity under primitive conditions. The extent to which the lives of these groups represent the lives of our species forty thousand years ago may likely never be known, but it may be the best we have to go on for now.

You may have noted that first births in modern hunter-gatherers happen a few years later than in wild chimpanzees, orangutans, elephants, and killer whales, all of which have first births four to five years earlier. Humans take a while to get going, reproductively. However, once we get started, we make up for lost time. The interval between births for humans, even food-stressed hunter-gatherers, is only three to four years, noticeably short compared to our closest primate relatives, such as chimpanzees (5.5 years) or orangutans (eight years). Helping one another with childcare apparently gives humans a reproductive advantage that may explain our ecological success compared to other apes—the fact that we have overrun the earth.

Before tackling the issue of how long these modern hunter-gatherers live, I want to reiterate something about the most common measure for characterizing human longevity—life expectancy. Life expectancy is an easy and convenient way to characterize the typical length of life in modern industrialized societies. However, that is a comparatively recent development. As commonly used, *life expectancy* means "life expectancy *at birth*," which is, in essence, the average ages at which people die. When infant and childhood mortality are high, that average will not be representative of adult longevity. For instance, in 1900 France, where birth and death records were excellent, life expectancy was forty-five years because of high infant and childhood mortality, but the most common age for adults to die was the early seventies.[9] In all of the small-scale, nontechnological populations studied, infant mortality is high. About 40 percent of children die by age ten in all three of the groups despite their ecological differences.

A better metric to characterize the longevity of adults, avoiding the childhood mortality issue, is life expectancy of those who survive to adulthood. Taking adulthood to start roughly at age fifteen, women in all three of our groups can expect to live another forty-three to forty-five years, meaning that adult lives average a total of about fifty-eight to sixty years in length. Again, putting these numbers in context, the same number (expected longevity of fifteen-year-old women) for the United States currently is another 67 years or eighty-two years in total. As with our animal data, estimates of the oldest individuals are the least reliable, but it appears that in all of these foraging societies, a few individuals seem to reach age eighty, and some may make it to their mid-eighties. This would make humans in the wild just a bit longer-lived than elephants and give them a longevity quotient *in the wild* of 3.8, better than other chimpanzees but nowhere near bat- or naked mole-rat like. We are, as most people no doubt assume, the longest-lived terrestrial mammal, even in a state of nature.

An important discovery about the health and longevity of these modern hunter-gatherers is that they avoid some of the major maladies of aging that strike modern humans living in the modern world. In particular, cardiovascular diseases and osteoporosis, two of our most common diseases of aging, are virtually unknown in hunter-gatherers, no doubt due to chronic high levels of physical activity and low-fat diets. All of these populations have exceptionally low cholesterol and blood pressure with virtually no calcification of their coronary arteries. Even in Papua New Guinea, which has the world's highest frequency of a gene variant (ApoE4) that in Western societies is associated with an increased risk for cardiovascular and Alzheimer's disease, the people living traditional foraging lives have essentially no cardiovascular disease. During my time there, I heard of a man from one of the Miyanmin villages who was fortunate enough to be hired by an Australian company to operate heavy equipment in the company's mining operation a short helicopter ride away. After several months training (he had never seen, much less operated, heavy equipment before) and living in company facilities, eating a traditional English/Australian diet rich in fatty domesticated beef, potatoes, and gravy, he dropped dead of a heart attack at age forty-five. His village sued the mining company. Never before had anyone in the village that age suddenly dropped dead for no apparent reason.

Researchers have noticed the exceptional bone density found in hunter-gatherers that protect against osteoporosis. This can even be seen in ancient skeletal remains of hunter-gatherers. One study from North America found that bone density in skeletons from seven thousand years ago, when indigenous people were still hunting and gathering, was 20 percent greater than farming people living in the same area some six thousand years later.[10] A similar difference in bone density has been found between modern pet dogs and wolves in the wild. I never saw a single older person in Papua New Guinea who was bent with age. They might have arthritic knees and hobble because of that, but their bones and spine remain remarkably sturdy. If you examine photos of older !Kung, Ache, or Hadza, you will notice the same thing.

MENOPAUSE

Human females typically have their last birth in their late thirties or early forties and become physiologically incapable of reproducing at about age fifty, with thirty years or so of their lives still ahead of them. The immediate reason is that their ovaries become depleted of eggs and they stop ovulating. Even in modern hunter-gatherer societies, women who reach that menopausal age can expect to live another decade or two. From an evolutionary perspective, maximizing reproduction is what nature favors, so why evolution allows women to run out of eggs when they still have decades to live is more than a little puzzling. So puzzling, in fact, that something of an academic cottage industry has developed around trying to understand it. Disagreements about whether and which other mammal species experience something similar to menopause abound. In fact, I attribute the disagreement about how long killer whales live to whether researchers believe—or not—that killer whales routinely experience something like menopause.

Some researchers are convinced that menopause has been favored by natural selection. Their logic goes like this. As women age, their survival horizon shortens, and the risk of a successful completed pregnancy decreases. Eventually, women reach an age where they can more successfully pass on their genes by devoting all their efforts to raising their last children or by helping those children raise *their* children—being helpful mothers or grandmothers, in other words. In fact, this "grandmothering" effect has been credited by some with our delayed reproductive maturity, the comparatively early age at which human babies are weaned, and our shortened interval between births relative to the other chimpanzees.[11] Mothers in all societies, including those of modern hunter-gatherers, help their last children as much as they can and help their adult daughters with childcare as much as possible. That childcare consists mainly of sharing food. There is no controversy about that. Darwinian logic dictates that women help their genetic descendants as much as they can. There is also little controversy that human females are unique among primates and highly unusual, if not unique, among mammals in living so long past the end of reproduction, even in the wild, so to speak. However, there *is* considerable controversy about whether in our distant evolutionary past, grandmothers were common enough and could provide enough help to more than compensate evolutionarily for halting their own personal reproduction.

I'm not convinced that this controversy is focused on the right puzzle, though. To me, the puzzle is not so much why women stop reproducing early. Menopause does not happen abruptly, like flipping a switch. It is just the final stage of reproductive aging, which happens exceptionally early in women relative to their longevity. The evolutionary puzzle, then, is not, Why menopause? It is why women's ovaries age so much faster than the rest of their body. Other organs—heart, lungs, muscles, kidneys—all age more slowly. Women, even among hunter-gatherers, start becoming less fertile as early as their late twenties. In the years before menopause, there is a period called *perimenopause* when women experience erratic menstrual cycles and failure to ovulate. Energy-stressed !Kung women have their last child at an average age of thirty-four years. So menopause is just the final stage of a decades-long period of progressively becoming less and less likely to conceive, more and more likely to miscarry, and more and more likely to have a problem pregnancy or an abnormal child.

Men, of course, age reproductively as well. Sperm count and quality drops, and erectile dysfunction increases. Older fathers are more likely to produce children with birth defects, too. Still, men's reproductive aging proceeds at a slower pace, more in line with how the rest of their body ages. There are numerous records of men in their seventies and eighties producing children. Charlie Chaplin and Mick Jagger, for instance, both had children at age seventy-three.

There is actually nothing unique about this sort of sex difference in reproductive aging or the difference between female reproductive aging and aging of the rest of the female body. We see it all the time in mice and rats. But here is the critical point. We see it only in the laboratory, not in the wild. Mice and rats have sufficient eggs and reproductive capacity to last their entire *natural* lives of several months, just not their entire *laboratory* lives of two to three years. Female lab mice, in fact, have a somewhat earlier "menopause" than humans compared to the total length of their laboratory lives. Their reproductive aging in the laboratory, like humans, is much faster than the aging of heart, muscle, lungs, or kidneys. Also, reproductive capability in male mice falls more in line with the aging of *their* bodies.

My own suspicion about human menopause is that it is not adaptive in the sense that it has been favored because of mothering or grandmothering. If that were true, I suspect nature would not slowly turn off the reproductive tap for more than a decade prior to menopause. Also, the side effects of menopause, such as hot flashes and sleep problems, do not seem like they should accompany an adaptive phase of life. I'm more inclined to attribute menopause to that late jump in human longevity that our ancient bones suggest happened about forty thousand years ago, when we became unequivocally longer-lived than Neanderthals and chimpanzees. Our ovarian aging may not yet have adapted to that. That interpretation is supported by the fact that egg depletion in human and chimpanzee ovaries proceeds at approximately the same rate.[12] It is just that humans for the past tens of thousands of years have lived longer than chimpanzees. That interpretation might also predict that the age of menopause should be gradually increasing as our reproductive biology catches up with our recent longevity. Evidence for a later age at menopause is equivocal, and even if it is increasing, it is doing so at a glacial pace. So the still unsolved puzzle to me is not why women undergo menopause but why their reproductive system ages so much more quickly than the rest of their body. Also, why don't men do something similar?

In all fairness, there is a compelling counterargument to my opinion about this, which admittedly may be colored by my familiarity with laboratory rodents. That counterargument is that uncontroversially, some mammal species other than humans have long female postreproductive lives in the wild. Several species of whales fall into this category. There is no indication that these whale species have had a leap in longevity over the past few thousand years. Also in most, although not all, of these whales, females live in multigenerational groups where helping their last offspring raise *their* calves is, in theory, possible. It isn't clear what the nature of that help might be. It also isn't clear whether similar troublesome side effects of postreproductive life—the hot flashes and insomnia—also occur in these whales. I would certainly like to know, though.

HOW LONG CAN HUMANS LIVE?

Unlike ancient humans, about whom we can make only educated guesses, we know to the day how long modern humans live under modern conditions. By "modern conditions," I mean at times and places where a government's record-keeping infrastructure became extensive and reliable enough to identify the birth and death dates of virtually every individual. That happened more recently than you might think. Sweden is the platinum standard. Its birth and death records are good from about 1750 and impeccable since 1860. In larger countries with sizable rural populations, it happened later, and in less technological countries, it hasn't happened yet. Birth registration in the United States did not become universal until 1933. Sometimes records are lost due to wars or natural disasters. Japan lost many of its birth records during World War II. Still, by now we have billions of individual birth and death records and can report with confidence the broad outlines of modern human longevity.

Those broad outlines tell us that longevity has increased dramatically in technologically developed countries since about the beginning of the Industrial Revolution. One study reported that life expectancy, the most common measure of population longevity, has increased in the world's longest-lived countries by two and a half years per decade since 1840. Recall, however, that life expectancy (which is a shorthand for life expectancy *at birth*) is misleading about the length of adult life wherever infant mortality is high. Infant mortality was high everywhere until recently. It began declining in the developed world only once we discovered the importance of sanitation, clean water, and uncontaminated food and developed childhood vaccinations. In Sweden, infant mortality today is more than fifty times lower than what it was in 1900. It is low enough in Sweden and other developed countries today that it has virtually no impact on life expectancy. But it did in 1900. Swedish life expectancy at birth in 1900 was only fifty-two years, but a fifteen-year-old could expect to live to age sixty-four. Today, expected longevity of Swedes and the rest of the developed world is virtually identical at birth and at age fifteen. That life expectancy, by the way, is around eighty years. On the low end of life expectancy among these countries, we have the United States, with its pre-COVID-19 life expectancy of seventy-nine years, and on the high end, we have Japan, where life expectancy is 84½ years.

Those broad outlines also tell us that women are better survivors than men. Women live longer than men at every time and in every place where good records are available. For instance, life expectancy in Iceland in the nineteenth century dropped to as low as eighteen years during a particularly pestilence-ridden year, and in parts of the 1970s and 1980s, Iceland was the longest-lived country in the world. But in every single year, regardless of whether times were good or bad or whether life expectancy was short or long, Icelandic women lived longer than men. Sometimes the difference is relatively small, a difference of a few percentage points, and sometimes it is large. Today in the United States, women live about 6 percent longer than men. In Russia in the early 1990s, women lived 20 percent longer than men. In a number of European countries today, the life-expectancy gap is shrinking, but in Japan, it is growing. At any age, women die at lower rates than men of all the top causes of death except one—Alzheimer's disease—and we have no idea why. Women survive better than men late in life, but they also survive better from birth to age five. Women survived the Irish potato famine and the 1918 flu pandemic better than men.[13] They are surviving the COVID-19 pandemic better, too. Even prenatally, they seem to be better survivors. For prematurely born babies, male sex is considered a mortality risk. This is one of the most robust and least understood features of human biology.

Yet for all our knowledge about modern human longevity, myths and misinformation abound, particularly about extreme longevity.

If you ask someone their age, you may or may not get an accurate answer. Among Hollywood celebrities, for instance, my experience is that you are exceptionally unlikely to get an accurate answer. This isn't because people don't know their age. It is because of a peculiar form of human vanity in which being younger is seen as somehow being better. That is true up to a certain age, and then vanity seems to reverse direction. Self-declared age is likely to become exaggerated. Men seem to be particularly prone to this species of vanity. It isn't difficult to understand why. Instead of being just another codger, you become a remarkable survivor. If you are over a hundred years old—that is, if you claim to be over a hundred years old—you may even become a local celebrity. Journalists may wish to interview you about the secret of your remarkable longevity. At least this was true until recently. We now have so many hundred-year-olds that you may need to make to 105 or 110 years to become celebrated.

Due to this strange form of vanity, centenarians have tended to become less common whenever and wherever birth and death records become more accurate. The United States is a prime example of this. The unreliable official records of 1850 report that the United States had eleven centenarians per hundred thousand people, but by 1910, the still unreliable records said that only four people per hundred thousand lived that long. In 1910, the same government records tell us even less reliably that African Americans, despite being poor and persecuted and despite having an average longevity some fifteen years lower than white Americans, were twenty times more likely to reach a hundred years of age than white Americans. Official government records from other countries are similarly unreliable. For instance, official records tell us that people living in Argentina, Bolivia, Bulgaria, Ireland, and Russia in 1900 were more likely to reach age one hundred than today's Japanese, living in the longest-lived country in the world.[14] So claims of exceptional longevity always need to be examined with care.

A question that has bedeviled humans for millennia is whether there is an upper limit to human life, and if there is, what is it? In a conversation I once had with the Cambridge University Egyptologist John Baines, I learned that in ancient Egypt, some five thousand years ago, that limit was assumed to be 110 years. The seventeenth-century English carved their opinion into stone. In Westminster Abbey—where the British crown their kings and queens and bury their most famous poets, politicians, and scientists—lies the grave of Thomas Parr, with his purported birth and death years of 1483 and 1635, a cool 152 years. Old Parr, as he was known, invented a colorful life story, which included bachelorhood until age eighty, an extramarital affair at one hundred, and a second marriage at 120 with a child soon after. Rubens and Van Dyck painted his portrait (figure 14.1). Today Parr has what might be the greatest tribute a man might have—a whiskey named after him.

Of course, Parr did not live 152 years. A final honor he received was to be autopsied by William Harvey, the most famous English physician of his day. The autopsy reports that his organs did not look exceptionally old. Harvey, who had conversed with him before he died, tells us that he had virtually no memory of his earlier life, including the names of royalty, the

Figure 14.1

Thomas Parr, painted by Anthony van Dyck, possibly from life. Van Dyck was in London at the time when Parr became famous for a brief time for his self-proclaimed age of 152 years. *Source*: Courtesy of Perkins School for the Blind Archives, Watertown, MA.

occurrence of wars, or the price of bread. In fact, the only evidence we have of his age is his own word and the fact that he looked old. It is easy to feel superior to those gullible Englishmen in the seventeenth century, but similarly bogus claims and similarly suspect evidence have been swallowed in more recent times. *Life* magazine published a story in 1966 about a village in Azerbaijan where people lived into their 160s. Harvard physician Alexander Leaf, before discovering he had been a victim of misinformation, reported men—men!—living into their 130s in Vilcabamba, a remote

Andean valley in southern Ecuador. In fact, the unifying theme of all the modern reports of remote regions with exceptionally long-lived people is that the key to long life is doing hard physical work, having a simple lifestyle with a supportive social network, being near few accessible medical facilities, and most important, being a man without a verifiable birth record.

What about real extraordinary longevity? Knowing that we need to be skeptical of extreme claims and that extreme claims, in fact, require extraordinary evidence, let's now consider Jeanne Calment, the oldest fully verified human.

Jeanne Calment was born in the small town of Arles in southern France on February 21, 1875 (figure 14.2).[15] In that year, Ulysses S. Grant had just begun serving his second term as president of the United States, the French composer Maurice Ravel was born, and an international agreement on the length of a meter and weight of a kilogram was signed in Paris by seventeen nations, including the United States. Mme. Calment came from a prominent family in Arles, a town of 25,000 at the time. Longevity ran in the family. Her father was a shipbuilder who lived to age ninety-three. Her mother died at eighty-six, and her older brother lived to age ninety-seven. She married her wealthy double first cousin (their paternal grandfathers were brothers and grandmothers were sisters), Fernand Calment, at age twenty-one, and so her married and maiden names were identical. Because of her family wealth, she never had to work, and her life was filled with servants and hobbies. She died in 1997 at the age of 122 years, 164 days, just a few days after I told a national television audience while on book tour that if they hurried, they could still shake the hand that shook the hand of Vincent van Gogh. Van Gogh spent sixteen months in Arles in 1888 and 1889, during which time he painted and drew more than three hundred works, including many of his most famous paintings, sliced off his left ear in a manic rage, and ran into young Jeanne at least several times. He was anything but famous at the time, except perhaps locally for his bizarre behavior. He had his severed ear delivered to a young prostitute. Mme. Calment remembered him as dirty, smelly, poorly dressed, and generally disagreeable.

I've called being publicly named the oldest living person the most dangerous job in the world. No one gets out alive, and one is usually killed on the job within a few months. Jeanne Calment was the world's oldest

Figure 14.2
Jeanne Calment at age 120. Already the world's oldest-known person ever at this age, she would live two more years.
Source: Photo by Michel Pisano, Arles, France.

person for more than six and a half years, during which time she became famous. Her fame began in 1988, when Arles celebrated the centenary of Van Gogh's time spent there. The city fathers were astounded to discover someone still alive who had met the artist. After that, she was on the radar of demographers who were obsessed with the very old. They were skeptical of her age because most people claiming to be over 110 years old (*super-centenarians*, as they are called) are, on further investigation, found to be not that old. They dug through her personal background in every conceivable way, uncovering more than thirty documents confirming her age. They fact-checked details about her early life that she mentioned during interviews. All evidence agreed. She really was that old. Incidentally, taking Jeanne Calment as our oldest representative, captive humans would have a longevity quotient of 5.5, not quite as high as that of a naked mole-rat.

She was a small woman, standing only 132 centimeters (four feet, six inches) late in life, weighing only forty kilograms (eighty-eight pounds), although she would have been somewhat taller and heavier when young.

Not surprisingly, she was in good health for most of her life. She rode a bicycle until she broke a leg in a fall at age one hundred but quickly and fully recovered without physical therapy of any sort. After another fall that broke her hip and elbow at age 115, she spent the rest of her life in a wheelchair. She was virtually deaf and blind in these later years, when she refused to wear hearing aids or have her cataracts removed. Doctors, journalists, and demographers had to shout questions in her ear, but she retained her wit, and some of her quotes (such as "I only have one wrinkle, and I'm sitting on it") have outlived her.

She had only one child, a daughter, Yvonne, who was never in robust health and died of pneumonia at age thirty-six. Yvonne's only child, Frédéric, died in a car accident, also at age thirty-six, so the longest-lived person ever has no surviving descendants. At age ninety, having no surviving heirs, she agreed to sell her apartment to a forty-seven-year-old lawyer, André-François Raffray, under what today we would call a reverse mortgage. He agreed to pay her 2,500 francs per month in return for inheriting her apartment after she died. Some twenty-nine years later, while she lived on, the lawyer died, never having spent a single day in that apartment despite paying for it for almost three decades.

Jeanne Calment is a classic outlier, somewhat who achieved something that no one else approaches. Some twenty-one years after her death, despite an explosive increase in the number of global centenarians and supercentenarians, no other person has been confirmed to live even 120 years. Only two other people have come very close: American Sarah Knauss, who reached her 119th birthday and then died in 1999, and one other, a Japanese woman named Kane Tanaka, who is still alive at age 119 as I write in March 2022. After these three, we know of eleven women who made it to age 117, including Lucile Randon (aka Soeur André), who survived COVID-19 at age 116. The oldest man known to date was Jiroemon Kimura, who lived a few days past his 116th birthday.[16]

Some readers may have heard that there are doubts about Jeanne Calment's extraordinary longevity. There aren't. The doubts came from two amateur Russian researchers who in 2019 received more publicity than they deserved from claiming that Calment's age was fraudulent. As the paper trail of her longevity is unassailable, they used some minor inconsistencies in stories she told in her last years of life to claim that it was actually Jeanne who died in 1934 rather than her daughter, Yvonne. Yvonne, in this rather bizarre fantasy, impersonated her mother for the next sixty-three years. If true, of course, this would have required complicity by Jeanne's older brother, François, by Yvonne's husband, and by Yvonne's seven-year-old son. It would also have required this prominent family's entire circle of friends and acquaintances, her household staff, and all the merchants with whom she dealt to fail to notice that the fifty-nine-year-old woman they knew had changed her appearance and grown twenty-three years younger overnight.

What are we to make of the fact that the world's oldest known person died more than twenty years ago and that despite the explosive increase in the number of centenarians since that time, no one has yet approached her age? One reasonable thought is that Jeanne Calment represents something of a limit to human longevity and that we shouldn't expect anyone else to surpass her or at least to surpass her by much because she represents the maximum time the human body can last, given the best longevity genes combined with the healthiest environment and perhaps a bit of luck thrown into the mix.

I sincerely hope this isn't true, as it could cost me—or at least my descendants—$1 billion. Yes, that's *billion* with a *B*. Let me explain in the next chapter.

15 METHUSELAH'S ZOO
MOVING FORWARD

Everywhere on earth, people are living longer than ever before—*on average*. The fastest-growing age group is centenarians, although living to a hundred years of age is still a rare accomplishment. Fewer than one person in a thousand lives that long, even in Japan, today's longest-lived country. Rare though they may be, the number of centenarians alive today has almost quadrupled since Jeanne Calment's death in 1997. But for all this increase, some twenty-four years after her death, no one has approached Jeanne Calment's longevity record. For that matter, no one has surpassed the 119-year longevity of Sarah Knauss. It is also difficult to ignore the fact that the rate of life expectancy increase in the world's longest-lived countries has slowed appreciably, even before we were blasted by COVID-19. Life expectancy in the United States, for instance, has not increased since 2015.

If you want to start a brawl at a demography convention, bring up the subject of a "limit" to human life. Is there a limit to life expectancy? Is there a longevity limit that no human will ever surpass? Either question will probably do for one demographer or another to throw the first punch.

In 1980, Stanford physician James Fries made a strange, somewhat optimistic, somewhat pessimistic prediction.[1] He claimed—and still claims, in fact—that the limit of life expectancy is about eighty-five years. That's the pessimistic part of his prediction. The optimistic part is that he also predicted that science will continue to find ways to keep us healthy longer, so that more and more of those eighty-five years will be spent in good health. The period of ill health that many suffer will be compressed into a smaller and smaller slice of time. The alternative is frightening. More and more people, living longer and longer and pushing up against a limit of human life, could require more and more medical help and could live more and more years in pain—demented and disabled. Some people might say that we are reaching toward that dystopian future today as health-care systems worldwide groan under the weight of care for the elderly.

A decade after Fries made this prediction, it was echoed by a group of professional demographers, most notably S. Jay Olshansky from the University of Illinois Chicago, who has been particularly vocal on the issue.[2] Olshansky also weighed in on the length of maximum life. He offered then and still thinks that no one is likely to surpass Jeanne Calment's longevity record by more than a few years—ever. Other demographers have been equally vociferous about their opinion that human life has no limit. They think that life expectancy will keep rising for the foreseeable future and that maximum longevity records will be broken again and again. One group has predicted that people born after the year 2000, which includes all the students I teach today, can expect to live a century or more.[3] For what it's worth, some forty years after Fries's prediction, Japan now has a life expectancy of 84½ years. The "limit" people can smirk about this. The no-limits crowd would be quick point out that Japanese life expectancy is being dragged down by those wimpy men. Japanese women have already surpassed the Fries limit. They can now expect to live 87½ years.

It was mainly due to my appreciation for the lessons nature could teach us about living healthy and living long that Olshansky and I made our $1 billion wager, which I'll describe shortly. Recall that nature—in the guise of certain animals such as birds, bats, and mole-rats—has repeatedly discovered how to deal with damaging free radicals much better than humans can. Other species (like elephants and whales) have developed dramatically better cancer resistance than humans. Still others, such as my beloved quahogs, have evolved ways to keep muscles strong and hearts beating for centuries. At some point, I am confident, the full armamentarium of the biomedical research enterprise will be deployed to study and eventually understand these lessons nature has to teach us about preserving and prolonging health.

The biochemist Leslie Orgel, who is famous for his research on the origin of life, was fond of pointing out something that should be obvious to all readers of this book by now. In fact, he pointed it out so often that it has become known as "Orgel's second rule"—to wit, *evolution is cleverer than you are*. What Orgel meant by his second rule, of course, was that evolution, with several billion years and billions of species with which to tinker, will have discovered solutions to problems that humans might never dream of. In the context of prolonging our health, this means that nature will have discovered many ways of combating the inherently destructive processes of life, such as free-radical damage and protein misfolding. Given that such a well-respected scientist pointed out such an obvious truth decades ago, I am somewhat astonished that the biomedical research community has stuck largely with studying animals that are so demonstrably failures at combating these processes. The workhorse of medical research continues to be the laboratory mouse—one of the shortest-lived and most cancer-prone mammals known. In a certain sense, I understand why. So much work has gone into developing tools for instructive intervening in mouse biology that we can do more sophisticated experiments with the mouse than any other mammal. We can deliberately turn individual genes on or off in any part of the mouse body at any time during a mouse's life. We can insert genes from humans, whales, bats, or other species into the mouse and turn them on and off when and where we wish. But genes do not operate in isolation. A whale gene in a mouse may do little more than caricature its role in its hometown, so to speak. Genes' activities must be coordinated like the instruments in an orchestra if you want them to produce beautiful music. Introducing a car horn into an orchestra is not likely to improve its music, no matter how useful the car horn may be in its native environment.

Because of the mouse's short life, we can also determine quickly whether a particular gene variant or new drug will preserve health or life in a mouse. In fact, researchers focusing on the biology of aging have already discovered about a dozen drugs that keep mice healthy and alive longer. Some of these drugs are in early human trials as I write. I purposely am not mentioning the names of any of them because some people are so desperate to live longer they might start taking them before we know for sure whether they are safe, much less effective, for people. What works in mice does not necessarily work in humans.

Certainly, some of these drugs may represent longevity breakthroughs. Time will tell. But remember, mice are losers in the game of healthy longevity. An exercise designed to improve the gait of the lame may be unlikely enhance the speed of an already accomplished sprinter. Mice are lame, but humans are already accomplished sprinters. So a drug that allows a mouse to live three rather than two years (or a fruit fly three rather than two months) may be unlikely to extend human health. Human biology may have already solved whatever problems limit a mouse's life. Don't forget, we are already the longest-lived terrestrial mammal. A mouse could learn a great deal about improving and extending its health from studying us. From this perspective, it is hardly surprising that only about one in ten cancer therapies effective in mice has turned out to also be effective in people. We are certainly grateful for the one in ten of those therapies, but might there be a more evolutionarily sensible approach to prolong health? For Alzheimer's disease, none of the over three hundred mouse successes has succeeded in people.

Medical research is as inherently tradition-bound and conservative as any ecclesiastical hierarchy. Funds for research are distributed according to the opinions of scientists who are exquisitely well trained in spotting flaws and detecting uncertainties in traditional experimental paradigms. I ought to know, as I have served on many, many such committees, and I plead guilty to having weighed in on such flaws and uncertainties as I found. There is nothing wrong with such scientific conservatism. It prevents money being wasted on hopelessly wrong-headed research.

But there is also a role for the scientifically adventurous and for out-of-normal-bounds research—for the wild and crazy idea that just might turn out to be true and, if so, then revolutionary. An acquaintance of mine, who also happens to be a Nobel Prize winner, likes to recount with glee how the work that won him his Nobel Prize was the only part of his research proposal that was rejected by a governmental review group.

But I think this hidebound approach to health research is changing. The bestiary of acceptable species on which respectable researchers can experiment is expanding. Naked mole-rats and blind mole-rats are now safely within the research bestiary. That progress may be due to another kind of limit—the limit of what we can learn from studying short-lived, cancer-prone laboratory species. As more and more people realize that nature provides us many examples of animals that combat fundamental aging processes more successfully than humans, there will be pressure to see what we can learn from those species. Some of that pressure may come from the private sector, where some very wealthy people appear to have a personal interest in remaining healthy longer. If you pay attention to headlines, this already seems to be happening.

We are not likely to have laboratory colonies of Greenland sharks, bowhead whales, rougheye rockfish, or even Brandt's bats any time soon. The good news is that while we may not have whales in the lab, we can have whales in a dish. That is, we can grow and study whale cells grown in the lab in exquisite detail today. The 2012 Nobel Prize in physiology or medicine was won by Shinya Yamanaka for discovering how to transform skin, liver, blood, or virtually any cell type into stem cells. Stem cells in a dish can in turn be transformed back into heart cells, muscle cells, or brain cells or even turned into miniature organs. An obvious use of the Yamanaka technology is to develop it to grow replacement parts for aging humans *from their own cells*. We are not far from being able to use this technology to cure certain diseases such as diabetes and Parkinson's disease. But a less obvious use of Yamanaka technology is to study how bird or bat or whale or shark brain or muscle cells deal with damaging free radicals and avoid turning cancerous or how quahog cells avoid misfolding their proteins for centuries.

Methuselah's Zoo, I believe, holds the key to prolonging human health. It may seem like a radical idea but perhaps a radical idea whose time has come. Let's all agree to acknowledge that *evolution is cleverer than you are*. Are you listening, Silicon Valley zillionaires?

It was this sort of thinking that led to my $1 billion wager.

It was 2001. I found myself sitting in a small conference room on the UCLA campus with perhaps a dozen scientists and a reporter from the *New York Times*. We had come together to discuss the future of human health. The reporter asked a question: when will we see the first 150-year-old human? We shifted uncomfortably in our seats. No one wanted to go out on a limb—except me. I blurted out, "I think that person is already alive." As I think back on that moment, it seems like that was exactly the right question to ask. And, amazingly, I think I gave exactly the right answer.

No one, I suspect, thinks that we will ever see a 150-year-old human, someone nearly thirty years older than Jeanne Calment, just because we have gotten better and better at diagnosing and treating individual diseases like cancer, stroke, and dementia. I certainly don't think that. It will happen only if we learn to treat aging itself as if it were a disease and delay or eliminate all those diseases simultaneously.

Jay Olshansky, premier public skeptic of exceptional longevity, whom I already knew and respected, read an account of this conference and phoned me to disagree. How strongly did I believe that, he asked. Would I like to make a friendly wager?

We didn't actually put up half a billion dollars each. Neither of our university salaries were quite up to that. What we did decide to do was put up $150 apiece. It had a nice symmetry. $150 each for 150 years to see if a 150-year-old human was alive. Olshansky did some quick back-of-the-envelope calculations. At the historic growth rate of the US stock market, our $300 could in 150 years turn into about $500 million. A dozen years later, when no one had still approached the age of Jeanne Calment, a reporter asked us once again whether we still felt confident that we would win our bet. We both did. To prove it, we doubled its size, each putting another $150 into the pot. Now we could safely claim that our wager was for a cool $1 billion. Even better, Olshansky had been actively investing our money, and now some twenty years after we made the wager, our pot had grown at considerably faster than the historical rate of the US stock market.

So what exactly was the wager? If by the year 2150 there exists or has ever existed a single, thoroughly documented 150-year-old and if that 150-year-old is mentally competent enough to hold a simple conversation, then my descendants—or in the best of all scenarios, I myself—will get the accumulated wealth. If not, then Olshansky's descendants will inherit the money.

I keep documentation of the wager in a safe place. My daughters have been informed of their—or their sons' and daughters'—future wealth. In many public debates and private conversations, Olshansky and I have discovered that we agree on many things. We agree that traditional medical research will not get us to the 150-year-old human. We agree that the only way to accomplish that is to find ways to treat aging itself as if it were a disease. A relatively small group of scientists, including yours truly, is working on exactly this in a new research specialty we call *geroscience*. Olshansky and I disagree only on how rapidly the big breakthroughs in treating aging will occur. Most of my geroscientist colleagues are sticking with the tried and true laboratory animals. But a few are now branching out. Many species with exceptional resistance to aging now have now had their genomes sequenced, and their cells are safely tucked away in laboratories, where researchers labor to learn their secrets. On the day that we can rely on staying healthy for ninety or a hundred years and somewhere someone is 150 years old or older, then we will have the creatures in Methuselah's Zoo to thank.

Appendix

Maximum recorded longevity (years) of selected species discussed in the text					
Species	Longevity (in years)	Wild (W) or captive (C)	Longevity quotient (LQ)	Precision	Age first offspring produced (in years)
Lab mouse	3	C	0.7	K	0.2
House sparrow	20	W	3.6	K	0.5–1
Major Mitchell's cockatoo	83	C	9.7	K	3–4
Laysan albatross	70	W	5.2	K	7–8
Manx shearwater	55	W	6.0	K	5–7
Wild turkey	15	W	1.0	K	1
Little brown bat	34	W	7.5	K	1
Common vampire bat	30	C	5.5	K	1
Common vampire bat	18	W	3.3	K	1
Indian flying fox	44	C	4.1	K	2
Brandt's bat	41	W	10.0	K	1
Giant tortoise	175	C	N/A	E	20–25
Tuatara	110	C	10.3	E	10–20
Naked mole-rat	39	C	6.7	E	Eusocial*
Blind mole-rat	21	C	2.9	K	1
Northern slimy salamander	20	C	5.3	K	3

(*continued*)

Maximum recorded longevity (years) of selected species discussed in the text
(continued)

Species	Longevity (in years)	Wild (W) or captive (C)	Longevity quotient (LQ)	Precision	Age first offspring produced (in years)
Olm	102	W	21.0	G	16
African elephant	74	W	1.6	E	11–14
Asian elephant	80	C/W	1.7	E	7–17
Chimpanzee	69	W	3.3	E	13
Orangutan	59	C	2.6	E	15
Capuchin	54	C	4.3	K	6–7
Human	86	W	3.8	E	19
Human	122	C	5.5	K	11–17
Lake sturgeon	152	W	8.5	E	15–25
Beluga sturgeon	118	W	Unknown	E	15–20
Rougheye rockfish	205	W	14.0	K	20
Greenland shark	392	W	11.7	G	156
Bottlenose dolphin	67	W	2.2	K	8
Killer whale	85	W	1.6	E	15
Blue whale	110	W	1.0	E	10
Fin whale	114	W	1.5	E	6–12
Bowhead whale	211	W	2.6	G	18–25

Notes: *The naked mold-rat is eusocial, so the time of first breed depends on the presence or absence of an already breeding queen. Age of first reproduction therefore is not defined. Precision: K = known from direct observation; E = estimated, accurate to within a few percentage points; G = guesstimate: a Bbest guess that is made using available techniques but that c. Could be wildly wrong, however. Age (years) first offspring produced.

Notes

CHAPTER 1

1. W. I. Lane and L. Comac, *Sharks Don't Get Cancer* (New York: Avery, 1992).

2. Lane and Comac, *Sharks Don't Get Cancer*.

3. S. L. Murphy, J. Xu, K. D. Kochanek, et al., "Mortality in the United States, 2017," *NCHS Data Brief* 328 (2018): 1–8.

4. S. N. Austad, "The Geroscience Hypothesis: Is It Possible to Change the Rate of Aging?," in *Advances in Geroscience*, ed. F. Sierra and R. Kohanski (New York: Springer, 2015), 1–36.

5. S. N. Austad and K. E. Fischer, "Mammalian Aging, Metabolism, and Ecology: Evidence from the Bats and Marsupials," *Journal of Gerontology* 46, no. 2 (1991): B47–B53.

CHAPTER 2

1. J. B. S. Haldane, *On Being the Right Size* (Oxford: Oxford University Press, 1985).

2. Mayflies are an exception that proves the rule. They are the only insects in which subadults have functional wings. The one- to two-day mayfly life stage (subimago) just preceding final molt does have wings, although it flies poorly.

3. F. Z. Molleman, B. J. Zwann, P. M. Brakefield, and J. R. Carey, "Extraordinary Long Life Spans in Fruit-Feeding Butterflies Can Provide Window on Evolution of Life Span and Aging," *Experimental Gerontology* 42, no. 6 (2007): 472–482.

CHAPTER 3

1. R. W. Coulson, J. D. Herbert, and T. D. Coulson, "Biochemistry and Physiology of Alligator Metabolism *in Vivo*," *American Zoologist* 29 (1989): 921–934.

2. G. M. Erickson, P. J. Makovicky, P. J. Currie, M. A. Norell, S. A. Yerby, and C. A. Brochu, "Gigantism and Comparative Life-History Parameters of Tyrannosaurid Dinosaurs," *Nature* 430 (2004): 772–775.

3. F. Rimblot-Baly, A. de Ricqlès, and L. Zylberberg, "Analyse paléohistologique d'une série de croissance par-tielle chez *Lapparentosaurus madagascariensis* (Jurassiquemoyen): Essai sur la dynamique de croissance d'undinosaure sauropode," *Annales de paléontologie* 81 (1995): 49–86.

CHAPTER 4

1. "European Longevity Records," Longevity List, Euring: Co-ordinating Bird Ringing throughout Europe, April 5, 2017, https://euring.org/data-and-codes/longevity-list.

2. F. Bacon, *The Historie of Life and Death* (Kessinger, 1638).

3. D. B. Botkin and R. S. Miller, "Mortality Rates and Survival of Birds," *American Naturalist* 108, no. 960 (1974): 181–192.

4. J. A. Clark, R. A. Robinson, D. E. Balmer, S. Y. Adams, M. P. Collier, M. J. Grantham, J. R. Blackburn, and B. M. Griffin, "Bird Ringing in Britain and Ireland in 2003," *Ringing and Migration* 22, no. 2 (2004): 85–127.

5. J. E. Cardoza, "A Possible Longevity Record for the Wild Turkey," *Journal of Field Ornithology* 66, no. 2 (1995): 267–269.

6. W. A. Calder and L. L. Calder, "Broad-Tailed Hummingbird: *Selasphorus platycercus*," in *The Birds of North America*, no. 16, ed. A. Poole, P. Stettenheim, and F. Gill (Philadelphia: American Ornithologists' Union, 1992), 1–16.

CHAPTER 5

1. W. H. Davis and H. B. Hitchcock, "A New Longevity Record for the Bat *Myotis lucifugus*," *Bat Research News* 36, no. 1 (1995): 1–6.

2. G. S. Wilkinson, "Vampire Bats," *Current Biology* 29, no. 23 (2019): R1216–R1217.

3. J. Maruthupandian and G. Marimuthu, "Cunnilingus Apparently Increases Duration of Copulation in the Indian Flying Fox, *Pteropus giganteus*," *PLOS ONE* 8, no. 3 (2013): e59743.

4. Beth Autin (associate director of library services) and Melody Brooks (registrar), San Diego Zoo, personal communication with the author, 2020.

5. A. J. Podlutsky, A. M. Khritankov, N. D. Ovodov, and S. N. Austad, "A New Field Record for Bat Longevity," *Journals of Gerontology A: Biological Science Medical Science* 60, no. 11 (2005): 1366–1368.

6. G. S. Wilkinson and D. M. Adams, "Recurrent Evolution of Extreme Longevity in Bats," *Biology Letters* 15, no. 4 (2019): 20180860.

7. P. Kortebein, B. Symons, A. Ferrando, D. Paddon-Jones, et al., "Functional Impact of 10 Days of Bed Rest in Healthy Older Adults," *Journals of Gerontology: Medical Sciences* 63A, no. 10 (2008): 1076–1081.

8. K. Lee, J. Y. Park, W. Yoo, T. Gwag T, et al., "Overcoming Muscle Atrophy in a Hibernating Mammal Despite Prolonged Disuse in Dormancy: Proteomic and Molecular Assessment," *Journal of Cellular Biochemistry* 104 (2008): 642–656.

9. D. D. Moreno Santillán, T. M. Lama, Y. T. Gutierrez Guerrero, et al., "Large-Scale Genome Sampling Reveals Unique Immunity and Metabolic Adaptions in Bats," *Molecular Ecology* (June 19, 2021), epub ahead of print.

10. D. Jebb, Z. Huang, M. Pippel, G. M. Hughes, et al., "Six Reference-Quality Genomes Reveal Evolution of Bat Adaptations," *Nature* 583, no. 7817 (2020): 578–584.

CHAPTER 6

1. J. D. Congdon, R. D. Nagleb, O. M. Kinney, R. C. van Loben Sels, et al., "Testing Hypotheses of Aging in Long-Lived Painted Turtles (*Chrysemys picta*)," *Experimental Gerontology* 38 (2003): 765–772.

2. L. Hazley, personal communication with the author, 2020.

CHAPTER 7

1. L. Keller, "Queen Lifespan and Colony Characteristics in Ants and Termites," *Insectes Sociaux* 45 (1998): 235–246.

2. K. D. Bozina, "How Long Does the Queen Live?," *Pchelovodstvo* 38 (1961): 13.

3. G. P. Slater, G. D. Yocum, and J. H. Bowsher, "Diet Quantity Influences Caste Determination in Honeybees," *Proceedings of the Royal Society B* 287 (2020): 20200614.

4. V. Chandra, I. Fetter-Pruneda, P. R. Oxley, A. L. Ritger, et al., "Social Regulation of Insulin Signaling and the Evolution of Eusociality in Ants," *Science* 361, no. 6400 (2018): 398–402.

CHAPTER 8

1. J. U. M. Jarvis, "Eusociality in a Mammal: Cooperative Breeding in Naked Mole-Rat Colonies," *Science* 212 (1981): 571–573.

2. Personal communication from Rochelle Buffenstein. The oldest animal reported in peer-reviewed literature was thirty-seven years old, but that animal was still alive—now thirty-nine years old—at the time of this writing.

3. S. Braude, S. Holtze, S. Begall, J. Brenmoehl, et al., "Surprisingly Long Survival of Premature Conclusions about Naked Mole-Rat Biology," *Biological Reviews of the Cambridge Philosophical Society* 96, no. 2 (2021): 376–393.

4. S. Liang, J. Mele, Y. Wu, R. Buffenstein, and P. J. Hornsby, "Resistance to Experimental Tumorigenesis in Cells of a Long-Lived Mammal, the Naked Mole-Rat (*Heterocephalus glaber*)," *Aging Cell* 9, no. 4 (2010): 626–635.

5. X. Tian, J. Azpurua, C. Hine, A. Vaidya, M. Myakishev-Rempel, et al., "High Molecular Weight Hyaluronan Mediates the Cancer Resistance of the Naked Mole-Rat," *Nature* 499, no. 7458 (2013): 346–349.

6. B. Andziak, T. P. O'Connor, Q. Wenbo, E. M. DeWall, et al., "High Oxidative Damage Levels in the Longest-Living Rodent, the Naked Mole-Rat," *Aging Cell* 5 (2006): 463–471.

7. I. Manov, M. Hirsh, T. C. Iancu, A. Malik, et al., "Pronounced Cancer Resistance in a Subterranean Rodent, the Blind Mole-Rat, *Spalax*: *In Vivo* and *in Vitro* Evidence," *BMC Biology* 11 (2013): 91.

8. V. Gorbunova, C. Hine, X. Tian, J. Ablaeva, et al., "Cancer Resistance in the Blind Mole Rat Is Mediated by Concerted Necrotic Cell Death Mechanism," *Proceedings of the National Academy of Sciences USA* 109, no. 47 (2021): 19392–19396.

9. D. R. Knight, D. V. Tappan, J. S. Bowman, H. J. O'Neill, and S. M. Gordon, "Submarine Atmospheres," *Toxicology Letters* 49 (1989): 243–251.

10. C. M. Ivy, R. J. Sprenger, N. C. Bennett, B. van Jaarsveld, et al., "The Hypoxia Tolerance of Eight Related African Mole-Rat Species Rivals That of Naked Mole-Rats, Despite Divergent Ventilator and Metabolic Strategies in Severe Hypoxia," *Acta Physiologica* 228, no. 4 (2020): e13436.

11. I. Shams, A. Avivi, and E. Nevo, "Oxygen and Carbon Dioxide Fluctuations in Burrows of Subterranean Blind Mole Rats Indicate Tolerance to Hypoxic-Hypercapnic Stresses," *Comparative Biochemistry and Physiology, Part A* 142 (2005): 376–382.

12. Y. Voituron, M. de Fraipont, J. Issartel, O. Guillaume, and J. Clobert, "Extreme Lifespan of the Human Fish (*Proteus anguinus*): A Challenge for Ageing Mechanisms," *Biology Letters* 7 (2011): 105–107.

13. J. Issartel, F. Hervat, M. de Fraipont, and Y. Voituron, "High Anoxia Tolerance in the Subterranean Salamander, *Proteus anguinus*, without Oxidative Stress nor Activation of Antioxidant Defenses during Reoxygenation," *Journal of Comparative Physiology B* 179 (2009): 543–551.

CHAPTER 9

1. I. McComb, G. Shannon, K. N. Sayialel, and C. Moss, "Elephants Can Determine Ethnicity, Gender, and Age from Acoustic Cues in Human Voices," *Proceedings of the National Academy of Sciences* 111, no. 14 (2014): 5433–5438.

2. L. J. West, C. M. Pierce, and W. D. Thomas, "Lysergic Acid Diethylamide: Its Effects on a Male Asiatic Elephant," *Science* 138, no. 3545 (1962): 1100–1103.

3. F. Thomas, R. Renaud, E. Benefice, T. De Meeüs, and J.-F. Guegan, "International Variability of Ages at Menarche and Menopause: Patterns and Main Determinants," *Human Biology* 73, no. 2 (2001): 271–290.

4. Much of my account of elephant life and use in the Burmese logging industry comes from a very nice unpublished PhD thesis by Khyne U. Mar, University College London, 2007.

5. R. Clubb, M. Rowcliffe, P. Lee, K. U. Mar, C. Moss, and G. J. Mason, "Compromised Survivorship in Zoo Elephants," *Science* 322 (2008): 1649.

6. The vast majority of information on African elephants is taken from various chapters in C. J. Moss, H. Croze, and P. C. Lee, eds., *The Amboseli Elephants* (Chicago: University of Chicago Press, 2011).

7. M. Sulak, L. Fong, K. Mika, S. Chigurupati, et al., "TP53 Copy Number Expansion Is Associated with the Evolution of Increased Body Size and an Enhanced DNA Damage Response in Elephants," *eLIFE* 5 (2016): e11994.

CHAPTER 10

1. S. N. Austad and K. E. Fischer, "Primate Longevity: Its Place in the Mammalian Scheme," *American Journal of Primatology* 28 (1992): 251–261.

2. S. Herculano-Houzel, *The Human Advantage: A New Understanding of How Our Brain Became Remarkable* (Cambridge, MA: MIT Press, 2016).

3. I should note that Herculano-Houzel has published data (not including bats) suggesting that in birds and mammals, a species' number of cortical neurons correlates with longevity, age at maturity, and length of postreproductive life. S. Herculano-Houzel, "Longevity and Sexual Maturity Vary across Species with Number of Cortical Neurons, and Humans Are No Exception," *Journal of Comparative Neurology* 527 (2019): 1689–1705. The meaning of this correlation, if any, would be greatly strengthened by measuring cell numbers of other organs to determine whether this is some special feature of the cortex.

4. K. Havercamp, K. Watanuk, M. Tomonaga, T. Matsuzawa, and S. Hirata, "Longevity and Mortality of Captive Chimpanzees from 1921 to 2018," *Primates* 60 (2019): 525–535.

5. H. Pontzer, D. A. Raichlen, R. W. Shumaker, C. Ocobock, and S. A. Wich, "Metabolic Adaptation for Low Energy Throughput in Orangutans," *Proceedings of the National Academy of Sciences USA* 107, no. 32 (2010): 14048–14052.

6. S. A. Wich, H. de Vries, M. Ancrenaz, L. Perkins, et al., "Orangutan Life History Variation," in *Orangutans: Geographic Variation in Behavioral Ecology and Conservation*, ed. S. A. Wich, S. S. Utami-Atmoko, T. Mitra Setia, and C. P. Van Schaik (Oxford: Oxford University Press, 2009), 65–75. This chapter summarizes the survival of zoo orangutans.

7. R. Weigl, *Longevity of Mammals in Captivity: From the Living Collections of the World* (Stuttgart: Schweizerbart, 2005). This book by Weigl is a compendium of zoo longevity records.

8. S. A. Wich, S. S. Utami-Atmoko, T. Mitra Setia, H. D. Rijksen, et al., "Life History of Wild Sumatran Orangutans (*Pongo abelii*)," *Journal of Human Evolution* 47 (2004): 385–398. This paper summarizes what is known about orangutan survival in the wild.

9. Weigl, *Longevity of Mammals in Captivity.*

10. S. A. Wich, R. W. Shumaker, L. Perkins, and H. De Vries, "Captive and Wild Orangutan (*Pongo* sp.) Survivorship: A Comparison and the Influence of Management," *American Journal of Primatology* 71 (2009): 680–686.

11. A. M. Bronikowski, J. Altmann, D. K. Brockman, M. Cords, et al., "Aging in the Natural World: Comparative Data Reveal Similar Mortality Patterns across Primates," *Science* 331 (2011): 1325–1328.

12. Weigl, *Longevity of Mammals in Captivity*.

13. H. Pontzer, D. A. Raichlen, A. D. Gordon, K. K. Schroepfer-Walker, et al., "Primate Energy Expenditure and Life History," *Proceedings of the National Academy of Sciences USA* 111,no. 4 (2014): 1433–1437.

14. Debbie Johnson, registrar, Brookfield Zoo, personal communication with the author, August 2021.

CHAPTER 11

1. T. A. Ebert and J. R. Southon, "Red Sea Urchins (*Strongylocentrotus franciscanus*) Can Live over 100 Years: Confirmation with A-Bomb Carbon," *Fisheries Bulletin* 101, no. 4 (2003): 915–922.

2. A. Bodnar and J. A. Coffman, "Maintenance of Somatic Tissue Regeneration with Age in Short- and Long-Lived Species of Sea Urchins," *Aging Cell* 15 (2016): 778–787.

3. P. G. Butler, A. D. Wanamaker Jr., J. D. Scourse, C. A. Richardson, and D. J. Reynolds, "Variability of Marine Climate on the North Icelandic Shelf in a 1,357-Year Proxy Archive Based on Growth Increments in the Bivalve *Arctica islandica*," *Palaeogeography, Palaeoclimatology, Palaeoecology* 373 (2013): 141–151.

4. M. A. Yonemitsu, R. M. Giersch, M. Polo-Prieto, M. Hammel, et al., "A Single Clonal Lineage of Transmissible Cancer Identified in Two Marine Mussel Species in South America and Europe," *eLIFE* 8 (2019): e47788.

5. M. Wisshak, M. López Correa, S. Gofas, C. Salas, et al., "Shell Architecture, Element Composition, and Stable Isotope Signature of the Giant Deep-Sea Oyster *Neopycnodonte zibrowii* sp. n. from the NE Atlantic," *Deep-Sea Research I* 56 (2009): 374–407.

6. Z. Ungvari, D. Sosnowska, J. B. Mason, H. Gruber, et al., "Resistance to Genotoxic Stresses in *Arctica islandica*, the Longest Living Noncolonial Animal: Is Extreme Longevity Associated with a Multi-stress Resistance Phenotype?," *Journals of Gerontology Biological Sciences & Medical Sciences* 68, no. 5 (2013): 521–529.

7. S. B. Treaster, A. Chaudhuri, and S. N. Austad, "Longevity and GAPDH Stability in Bivalves and Mammals: A Convenient Marker for Comparative Gerontology and Proteostasis," *PLoS One* 10, no. 11 (2015): e0143680.

CHAPTER 12

1. V. G. Carrete and J. J. Wiens, "Why Are There So Few Fish in the Sea?," *Proceedings of the Royal Society London B* 279 (2012): 2323–2329.

2. R. M. Bruch, S. E. Campana, S. L. Davis-Foust, M. J. Hansen, and J. Janssen, "Lake Sturgeon Age Validation Using Bomb Radiocarbon and Known-Age Fish," *Transactions of the American Fisheries Society* 138 (2009): 361–372.

3. "152-Year-Old Lake Sturgeon Caught in Ontario," *Commercial Fisheries Review* 6, no. 9 (1954): 28.

4. G. I. Ruban, and R. P. Khodorevskaya, "Caspian Sea Sturgeon Fisher: A Historic Overview," *Journal of Applied Ichthyology*27 (2011): 199–208.

5. G. M. Cailliet, A. H. Andrews, E. J. Burton, D. L. Watters, D. E. Kline, and L. A. Ferry-Graham, "Age Determination and Validation of Studies of Marine Fishes: Do Deep-Dwellers Live Longer?," *Experimental Gerontology* 36 (2001): 739–764.

6. S. R. R. Kolora, G. L. Owens, J. M. Vazquez, A. Stubbs, et al., "Origins and Evolution of Extreme Life Span in Pacific Ocean Rockfishes," *Science* 374 (2021): 842.

7. C. R. McClain, M. A. Balk, M. C. Behfield, T. A. Branh, et al., "Sizing Ocean Giants: Patterns of Intraspecific Size Variation in Marine Megafauna," *PeerJ* (2015): e715.

8. J. J. L. Long, M. G. Meekan, H. H. Hsu, L. P. Fanning, and S. E. Campana, "Annual Bands in Vertebrae Validated by Bomb Radiocarbon Assays Provide Estimates of Age and Growth of Whale Sharks," *Frontiers in Marine Science* 7 (2020): 188.

9. L. L. Hamady, L. J. Natanson, G. B. Skomal, and S. R. Thorrold, "Vertebral Bomb Radiocarbon Suggests Extreme Longevity in White Sharks," *PLOS ONE* 9, no. 1 (2014): e84006.

10. L. J. Natanson and G. B. Skomal, "Age and Growth of the White Shark, *Carcharodon carcharias*, in the Western North Atlantic Ocean," *Marine & Freshwater Research* 66, no. 5 (2015): 387–398.

11. Y. Y. Watanabe, N. L. Payne, J. M. Semmens, A. Fox, and C. Huveneers, "Swimming Strategies and Energetics of Endothermic White Sharks during Foraging," *Journal of Experimental Biology* 222 (2019): jeb185603.

12. Y. Y. Watanabe, C. Lydersen, A. T. Fisk, and K. M. Kovacs, "The Slowest Fish: Swim Speed and Tail-Beat Frequency of Greenland Sharks," *Journal of Experimental Marine Biology and Ecology* 426–427 (2012): 5–11.

13. S. Studenski, S. Perera, K. Patel, C. Rosano, et al., "Gait Speed and Survival in Older Adults," *Journal of the American Medical Association* 305, no. 1 (2011): 50–58.

14. J. Nielsen, R. B. Hedehohn, J. Heinemeier, P. G. Bushnell, et al., "Eye Lens Radiocarbon Reveals Centuries of Longevity in the Greenland Shark (*Somniosus microcephalus*)," *Science* 353 (2016): 702–704.

CHAPTER 13

1. R. S. Wells, "Social Structure and Life History of Bottlenose Dolphins near Sarasota Bay, Florida: Insights from Four Decades and Five Generations," in *Primates and Cetaceans: Field Research and Conservation of Complex Mammalian Societies*, ed. J. Uamagiwa and L. Karczmarski, Primatology Monographs (Kyoto: Springer Japan, 2014).

2. R. Wells, personal communication with the author, 2020.

3. R. C. Connor, *Dolphin Politics in Shark Bay: Journey of Discovery* (New Bedford, MA: Dolphin Alliance Project, 2018).

4. C. Kamiski, E. Kryszczyk, and J. Mann, "Senescence Impacts Reproduction and Maternal Investment in Bottlenose Dolphins," *Proceedings of the Royal Society B* 285 (2018): 20181123.

5. P. K. Olesiuk, M. A. Bigg, and G. M. Ellis, "Life History and Population Dynamics of Resident Killer Whales (*Orcinus orca*) in the Coastal Waters of British Columbia and Washington State," *Report of the International Whaling Commission*, special issue 12 (1990): 209–244.

6. E. Mitchell, and A. N. Baker, "Age of Reputedly Old Killer Whale, *Orcinus orca*, 'Old Tom' from Eden, Twofold Bay Australia," *Report of the International Whaling Commission*, special issue 3 (1980): 143–154.

7. T. R. Robeck, K. Willis, M. R. Scarpuzzi, and J. K. O'Brien, "Comparisons of Life-History Parameters between Free-Ranging and Captive Killer Whale (*Orcinus orca*) Populations for Application toward Species Management," *Journal of Mammalogy* 96, no. 5 (2015): 1055–1070.

8. J. Jett and J. Ventre, "Captive Killer Whale (*Orcinus orca*) Survival," *Marine Mammal Science* 31, no. 4 (2015): 1362–1377. This paper by a former killer whale trainer and a physician attempted to undercut Robeck and colleagues' claim that captive killer whales survived as well as wild killer whales did, leading to a rebuttal by Robeck and colleagues that effectively dismantled the data and analysis used by Jett and Ventre. T. R. Robeck, K. Jaakkola, G. Stafford, and K. Willis, "Killer Whale (*Orcinus orca*) Survivorship in Captivity: A Critique of Jett and Ventre," *Marine Mammal Science* 32, no. 2 (2016): 786–792. Field biologists interested in the postreproductive life of killer whales, analogizing it to human menopause, also weighed in on the greater longevity of their wild whales, also provoking a response from Robeck and colleagues. T. R. Robeck, K. Willis, M. R. Scarpuzzi, and J. K. O'Brien, "Survivorship Pattern Inaccuracies and Inappropriate Anthropomorphism in Scholarly Pursuits of Killer Whale (*Orcinus orca*) Life History: A Response to Franks et al. (2016)," *Journal of Mammalogy* 97, no. 3 (2016): 899–909. The feud caught the attention of the wider scientific community when it was highlighted in one of the highest-profile science journals in the world. E. Callaway, "Clash over Killer-Whale Captivity," *Nature* 531 (2016): 426–427.

9. F. L. Read, A. A. Hohn, and C. H. Lockyer, "A Review of Age Estimation Methods in Marine Mammals with Special Reference to Monodontids," *NAMMCO Scientific Publications* (2018): 10, https://doi.org/10.7557/3.4474.

10. Read, Hohn, and Lockyer, "A Review of Age Estimation Methods in Marine Mammals with Special Reference to Monodontids."

11. J. C. George, and J. R. Bockstoce, "Two Historical Weapon Fragments as an Aid to Estimating the Longevity and Movements of Bowhead Whales," *Polar Biology* 31 (2008): 751–754. This paper nicely summarizes the history of estimating age in bowhead whales.

12. J. C. George, J. Bada, J. Zeh, L. Scott, et al., "Age and Growth Estimates of Bowhead Whales (*Balaena mysticetus*) via Aspartic Acid Racemization," *Canadian Journal of Zoology* 77 (1999): 571–580.

13. M. S. Savoca, M. F. Czapanskiy, S. R. Kahane-Rapport, W. T. Gough, et al., "Baleen Whale Prey Consumption Based on High-Resolution Foraging Measurements. *Nature* 599, no. 7883 (2021): 85–90.

14. D. Tejada-Martinez, J. P. de Magalhães, and J. C. Opazo, "Positive Selection and Gene Duplications in Tumour-Suppressor Genes Reveal Clues about How Cetaceans Resist Cancer," *Proceedings of the Royal Society B* 288, no. 1945 (2021): 20202592.

CHAPTER 14

1. A. Bergström, C. Stringer, M. Hajdinjak, E. M. Scerri, and P Skoglund, "Origins of Modern Human Ancestry," *Nature* 590 (2021): 229–237.

2. E. Trinkaus, "Neanderthal Mortality Patterns," *Journal of Archaeological Science* 22 (1995): 121–142.

3. R. Caspari and S.-H. Lee, "Older Age Becomes Common Late in Human Evolution," *Proceedings of the National Academy of Sciences USA* 101, no. 30 (2004): 10895–10900.

4. J. Oeppen and J. W. Vaupel, "Broken Limits to Life Expectancy," *Science* 296 (2002): 1029–1031.

5. C. O. Lovejoy, R. S. Meindl, T. R. Pryzbeck, T. S. Barton, K. G. Heiple, and D. Kotting, "Paleodemography of the Libben Site, Ottawa County, Ohio," *Science* 198, no. 4314 (1977): 291–293.

6. A. Koch, C. Brierley, M. M. Maslin, and S. L. Lewis, "Earth System Impacts of the European Arrival and Great Dying in the Americas after 1492," *Quaternary Science Reviews* 207, no. 1 (2019): 13–36.

7. My account of the !Kung is taken largely from Nancy Howell's book *Demography of the Dobe Area !Kung* (New York: Academic Press, 1979). Similarly for the Ache, most of my information is from the book by Kim Kill and A. Magdalena Hurtado, *Ache Life History* (Hawthorn, NY: Aldine De Gruyter, 1996), and for the Hadza, from Nicholas Blurton Jones's book *Demography and Evolutionary Ecology of the Hadza Hunter-Gatherers* (Cambridge: Cambridge University Press, 2016).

8. J. P. Hurd, "The Shape of High Fertility in a Traditional Mennonite Population," *Annals of Human Biology* 33, no. 5/6 (2006): 557–569.

9. Details of modern and documented historical demography in this book are all taken from the outstanding Human Mortality Database maintained by the University of California, Berkeley (USA), and the Max Planck Institute for Demographic Research (Germany), and they are available at www.mortality.org or www.humanmortality.de.

10. T. M. Ryan and C. N. Shaw, "Gracility of the Modern *Homo sapiens* Skeleton Is the Result of Decreased Biomechanical Loading," *Proceedings of the National Academy of Sciences USA* 112, no. 2 (2015): 372–377.

11. K. Hawkes, J. F. O'Connell, N. G. Jones, H. Alvarez, and E. L. Charnov, "Grandmothering, Menopause, and the Evolution of Human Life Histories," *Proceedings of the National Academy of Sciences USA* 95, no. 3 (1998): 1336–1339.

12. K. P. Jones, L. C. Walker, D. Anderson, A. Lacreuse, S. L. Robson, and K. Hawkes, "Depletion of Ovarian Follicles with Age in Chimpanzees: Similarities to Humans," *Biology of Reproduction* 77 (2007): 247–251.

13. V. Zarulli, J. A. Barthold Jones, A. Oksuzyan, R. Lindahl-Jacobsen, K. Christensen, and J. W. Vaupel, "Women Live Longer Than Men Even during Severe Famines and Epidemics," *Proceedings of the National Academy of Sciences USA* 115, no. 4 (2018): E832–E840.

14. I discuss this issue more extensively in my book *Why We Age: What Science Is Discovering about the Body's Journey through Life* (New York: Wiley, 1997).

15. The life of Jeanne Calment is wonderfully told in the book *Jeanne Calment: From Van Gogh's Time to Ours* by Michel Allard, Victor Lèbre, and John-Marie Robine (New York: Freeman, 1998).

16. The world's oldest people (those who have lived at least 110 years) are age-validated and monitored by several groups of people interested in extreme longevity. One such group even keeps a Wikipedia page with lists. "Oldest People," Wikipedia, https://en.wikipedia.org/wiki/Oldest_people.

CHAPTER 15

1. J. F. Fries, "Aging, Natural Death, and the Compression of Morbidity," *New England Journal of Medicine* 303, no. 3 (1980): 130–135.

2. S. J. Olshansky, B. A. Carnes, and C. Cassel, "In Search of Methuselah: Estimating the Upper Limits to Human Longevity," *Science* 250 (1990): 634–640.

3. K. Christensen, G. Doblhammer, R. Rau, and J. W. Vaupel, "Ageing Populations: The Challenges Ahead," *The Lancet* 374 (2009): 1196–1208.

Further Reading

CHAPTER 1: DOCTOR DUNNET'S FULMAR

Austad, Steven N. *Why We Age*. New York: Wiley, 1997.

Calder, William A. *Size, Function, and Life History*. New York: Dover, 1986.

Clutton-Brock, T. H., ed. *Reproductive Success*. Chicago: University of Chicago Press, 1988.

Skloot, Rebecca. *The Immortal Life of Henrietta Lacks*. New York: Crown, 2011.

Weinberg, Robert A. *One Renegade Cell*. New York: Basic Books, 1999.

CHAPTER 2: INSECTS, SIZE, AND THE HISTORY OF LIFE

Dudley, Robert. *The Biomechanics of Insect Flight*. Princeton, NJ: Princeton University Press, 2000.

Fortey, Richard. *Life: A Natural History of the First Four Billion Years of Life on Earth*. New York: Knopf, 1998.

Haldane, J. B. S. *On Being the Right Size*. Oxford: Oxford University Press, 1985.

CHAPTER 3: PTEROSAURS

Witton, Mark P. *Pterosaurs*. Princeton, NJ: Princeton University Press, 2013.

CHAPTER 4: BIRDS

Attenborough, David. *The Life of Birds*. Princeton, NJ: Princeton University Press, 1998.

Lovette, Irby J., and John W. Fitzpatrick, eds. *Handbook of Bird Biology*. 3rd ed. New York: Wiley, 2016.

Ricklefs, Robert E., and Caleb E. Finch. *Aging: A Natural History*. New York: Scientific American Library, 1995.

CHAPTER 5: BATS

Griffin, Donald R. *Listening in the Dark*. New Haven, CT: Yale University Press, 1958.

Nowak, Ronald M. *Walker's Bats of the World*. Baltimore, MD: Johns Hopkins Press, 1994.

Tuttle, Merlin. *The Secret Lives of Bats*. New York: Houghton Mifflin Harcourt, 2015.

CHAPTER 6: ISLANDS AND REPTILES

Chambers, Paul. *A Sheltered Life: The Unexpected History of the Giant Tortoises*. Oxford: Oxford University Press, 2005.

Darwin, Charles. *The Voyage of the Beagle*. Originally published in 1839, now available in many versions.

Quammen, David. *The Song of the Dodo*. New York: Scribner, 1997.

Weiner, Jonathan. *The Beak of the Finch*. New York. Vintage, 1995.

CHAPTER 7: SOCIAL INSECTS

Bignell, David E., Yves Roisin, and Nathan Lo, eds. *The Biology of Termites: A Modern Synthesis*. Dordrecht: Springer, 2011.

Hölldobler, Bert, and Edward. O. Wilson. *The Ants*. Cambridge, MA: Belknap Press, 1990.

Tschinkel, Walter R. *The Fire Ants*. Cambridge, MA: Belknap Press, 2006.

CHAPTER 8: TUNNELS AND CAVES

Kelly, Scott. *Endurance: A Year in Space, a Lifetime of Discovery*. New York: Knopf, 2017.

Sherman, Paul W., J. U. M. Jarvis, and R. D. Alexander, eds. *The Biology of the Naked Mole-Rat*. Princeton, NJ: Princeton University Press, 1991.

CHAPTER 9: ELEPHANTS

Haynes, Gary. *Mammoths, Mastodons, and Elephants: Biology, Behavior and the Fossil Record*. Cambridge: Cambridge University Press, 1991.

Moss, Cynthia. J., Harvey Croze, and Phyllis C. Lee, eds. *The Amboseli Elephants*. Chicago: University of Chicago Press, 2011.

CHAPTER 10: NONHUMAN PRIMATES

Darwin, Charles. *The Descent of Man and Selection in Relation to Sex*. London: John Murray, 1871.

De Waal, Frans. *Are We Smart Enough to Know How Smart Animals Are?* New York: Norton, 2016.

Gould, Stephen. J. *The Mismeasure of Man*. New York: Norton, 1981. Excellent book on the misuses of brain size.

Herculano-Houzel, Suzana. *The Human Advantage*. Cambridge, MA: MIT Press, 2017. Describes her pioneering research on neuron numbers in various brain regions among diverse species.

Jerison, Harry J. *Evolution of the Brain and Intelligence*. New York: Academic Press, 1973. Invention of the encephalization quotient.

Pontzer, Herman. *Burn*. New York: Avery, 2021.

Weigl, Richard. *Longevity of Mammals in Captivity: From the Living Collections of the World*. Stuttgart: Schweizerbart, 2005.

Wich, Serge A., S. Suci Utami-Atmoko, Tatang Mitra Setia, and Carel P. Van Schaik, eds. *Orangutans: Geographic Variation in Behavioral Ecology and Conservation*. Oxford: Oxford University Press, 2009.

CHAPTER 11: URCHINS, WORMS, AND QUAHOGS

Gosling, Elizabeth. *Marine Bivalve Molluscs*. 2nd ed. New York: Wiley-Blackwell, 2015.

Nouvian, Claire. *The Deep: The Extraordinary Creatures of the Abyss*. Chicago: University of Chicago Press, 2007.

Rozwadowski, Helen M. *Vast Expanses: A History of the Oceans*. London: Reaktion Books, 2019.

CHAPTER 12: FISHES AND SHARKS

Abel, Daniel C., R. Dean Grubbs, and Elise Pullen. *Shark Biology and Conservation*. Baltimore: Johns Hopkins University Press, 2020.

Dipper, Frances and Mark Carwardine. *The Marine World: A Natural History of Ocean Life*. Ithaca, NY: Cornell University Press, 2016.

Shubin, Neil. *Your Inner Fish: A Journey into the 3.5-Billion-Year History of the Human Body*. New York: Vintage, 2009.

CHAPTER 13: WHALES

Connor, Richard C. *Dolphin Politics in Shark Bay: Journey of Discovery*. New Bedford, MA: Dolphin Alliance Project, 2018.

Eisenberg, John. F. *The Mammalian Radiations*. Chicago: University of Chicago Press, 1981.

George, J. C., and J. G. M. Thewissen. *The Bowhead Whale: Biology and Human Interactions*. Cambridge, MA: Academic Press, 2020.

Mann, Janet, Richard C. Connor, Peter L. Tyack, and Hal Whitehead, eds. *Cetacean Societies: Field Studies of Dolphins and Whales*. Chicago: University of Chicago Press, 2000.

Reynolds III, John E., Randall S. Wells, and Samantha D. Eide. *The Bottlenose Dolphin: Biology and Conservation*. Gainesville: University Press of Florida, 2013.

Shields, Monica Wieland. *Endangered Orcas: The Story of the Southern Residents*. Seattle: Orca Watcher, 2019.

Würsig, Bernd, J. G. M. Thewissen, and Kit M. Kovacs, eds. *Encyclopedia of Marine Mammals*. 3rd ed. Cambridge, MA: Academic Press, 2017.

CHAPTER 14: THE HUMAN LONGEVITY STORY

Allard, Michel, Victor Lèbre, and Jean-Marie. Robine. *Jeanne Calment: From Van Gogh's Time to Ours*. New York: Freeman, 1998.

Austad, Steven N. *Why We Age: What Science Is Discovering about the Body's Journey through Life*. New York: Wiley, 1997.

Blurton Jones, Nicholas. *Demography and Evolutionary Ecology of Hadza Hunter-Gatherers*. Cambridge: Cambridge University Press, 2016.

Hawkes, Kristin, and Richard R. Paine, eds. *The Evolution of Human Life History*. Santa Fe, NM: School of American Research Press, 2006.

Hill, Kim H., and A. Magdalena Hurtado. *Ache Life History: The Ecology and Demography of a Foraging People*. New York: Aldine de Gruyter, 1996.

Howell, Nancy. *Demography of the Dobe !Kung*. 2nd ed. Hawthorne, NY: Aldine de Gruyter, 2000.

Morren, George E. B. Jr. *The Miyanmin: Human Ecology in a Papua New Guinea Society*. Iowa City: Iowa State University Press, 1986.

Trinkaus, Erik, and Pat Shipman. *The Neandertals: Changing the Image of Mankind*. New York: Knopf, 1992.

Wrangham, Richard. *Catching Fire: How Cooking Made Us Human*. New York: Basic Books, 2009.

CHAPTER 15: METHUSELAH'S ZOO MOVING FORWARD

There are armfuls of books on the emerging science of living longer. Here are only a few of my favorite recent ones.

Armstong, Sue. *Borrowed Time: The Science of How and Why We Age*. London: Bloomsbury Sigma, 2019.

Barzilai, Nir, and Toni Robino. *Age Later*. New York: St. Martin's Press, 2020.

Gifford, Bill. *Spring Chicken: Stay Young Forever (or Die Trying)*. New York: Grand Central Publishing, 2015.

Sinclair, David A., and Matthew D. LaPlante. *Lifespan: Why We Age—and Why We Don't Have To*. New York: Atria Books, 2019.

Index

Note: Page numbers in *italics* indicate tables.

Longevity (cont.)
105, 109, 119–121, 124, 126–127,
157–158, 182–183, 192, 194–195, 201,
203–204, 206, 210, 212, 220, 224,
238–239, 241
kit, 27, 43, 107, 110
limit to, 142, 144, 223, 262, 267, 269–270
quotient (LQ), 12, 51–53, 67, 93, 117,
121, 148–149, 163, 169, 213, 224, *277*
relative, 11
sex differences in, 76, 260–262
trees, 8–10
Lophochroa leadbeateri. See Cockatoos, Major
Mitchell's
LQ. *See* Longevity, quotient
LSD, 135

Malaysia, 68
Manhattan Project, 57, 77
Mastotermes darwiniensis. See Termites, giant
northern
Maturity, sexual, 38–39, 53, 63, 70, 108,
136, 138, 166, 225, 228, 257. *See also*
Reproduction
Mayflies, 25, 27, 29, 279n
McCarthy, Colin, 92
Menopause, 138, 235, 256–259
Metabolism, 34, 55, 74–75, 93, 98–99,
112–113, 117, 119, 122, 135, 165, 173,
179, 181, 188, 194, 208, 214, 226–227,
236
rate of, 11, 54–55, 73, 97, 108, 113, 119,
123, 125, 173, 179, 180, 187, 188,
194–195, 206, 208, 214, 215, 226–227,
240
Methuselah, 7, 71
Methuselah's Zoo, 7, 11, 37, 99, 241, 273,
275
Midway atoll, 48–50, 83
Miller, Richard S., 46, 48
Millipedes, 20
Mitchell, Edward, 230–231

Mitochondria, 24, 37, 76
Miyanmin people, 250–253, 256
Moas (birds), 84
Mole-rats
Middle East blind, 120–121, 123, 146,
272, *277*
naked, 11, 114–120, 123, 125, 151, 172,
174, 256, 265, 272, *277*, 281n
Mollusks, 8, 188–198
Monkeys, 11, 147, 150–151, 170, 222, 224,
250
capuchin, 169–173, *278*
muriqui, 170, 171
owl, three-striped, 71
woolly spider (*see* Monkeys, muriqui)
Morren, George, 250–251
Mortensen, Hans Christian, 46–47
Mouse, 3, 6, 10–12, 21, 42, 55, 63, 64, 66,
84, 116, 117, 119, 146, 148, 150, 151,
172, 216
house, 3, 6, 42, *277*
laboratory, 3, 271–272, *277*
Muscles, 6, 11, 23–24, 33–34, 36–37,
53–54, 75–76, 131, 193, 258, 270
Mutations, 4, 22–23, 36, 68, 145
Myotis
brandtii (*see* Bats, Brandt's)
lucifugus (*see* Bats, little brown)

Natural selection, 23, 210, 257
Neanderthal humans, 144, 246–248,
259
Neuron number, 152–154, 156
Nipah virus, 62, 68
NMR. *See* Mole-rats, naked

Ocean, size and origin, 177–180
Octopus, 133, 189
Old Tjikko (tree), 8–10, 182
Old Tom (killer whale), 229–231
Olm, 124–127, *278*
Olshansky, S. Jay, 270, 274–275